The first edition of *Hubble Vision* caused a sensation. It became an international bestseller and won world-wide, critical acclaim – for both its beautiful illustrations and clear, precise, and lively text. This eagerly awaited second edition is the most comprehensive, most authoritative, and best illustrated popular book available on the Hubble Space Telescope. It provides a magnificent portfolio of the latest and greatest images from the HST, woven together with a lucid text explaining the most exciting discoveries and setting them in the context of our current understanding in astronomy.

This second edition has been completely revised, updated, and expanded to include all the latest astronomical discoveries – from supernovae and protostars to gravitational lensing, black holes, and the early universe. It is now even better illustrated, with more than 100 new figures. Throughout, the text provides sufficient scientific background for any reader to understand and appreciate the remarkable discoveries being made by this, the foremost observatory of our age. The unique combination of authors – an award-winning science writer and a key scientist involved in the development of the mission – ensure that the text is both engaging and authoritative.

Hubble Vision offers a view of the universe as never seen before. It will capture the imagination of all those interested in the astronomical quest of understanding our universe – from high-school students to general readers and amateur astronomers.

Carolyn Collins Petersen is a science journalist and creator of astronomy educational materials. She currently serves as books and products editor for Sky Publishing Corporation, and is editor of the popular annual, *SkyWatch Magazine*.

John C. Brandt is a senior research associate with the Laboratory of Atmospheric and Space Physics at the University of Colorado.

HUBBLE VISION

Further Adventures with the
Hubble Space Telescope

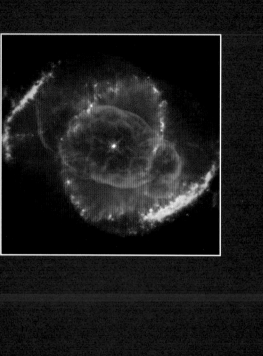

SECOND EDITION

HUBBLE VISION

Further Adventures with the Hubble Space Telescope

CAROLYN COLLINS PETERSEN

AND JOHN C. BRANDT

CAMBRIDGE
UNIVERSITY PRESS

PUBLISHED BY THE PRESS SYNDICATE OF THE UNIVERSITY OF CAMBRIDGE

The Pitt Building, Trumpington Street, Cambridge CB2 1RP, United Kingdom

CAMBRIDGE UNIVERSITY PRESS

The Edinburgh Building, Cambridge CB2 2RU, United Kingdom

40 West 20th Street, New York, NY 10011–4211, USA

10 Stamford Road, Oakleigh, Melbourne 3166, Australia

First published 1995
Reprinted 1996
Second edition 1998

Printed in the United Kingdom at the University Press, Cambridge

Typeset in Monotype Dante 10/13pt, in QuarkXpress™ [DS]

A catalogue record for this book is available from the British Library

Library of Congress Cataloguing in Publication data
Petersen, Carolyn Collins.
    Hubble vision: astronomy with the Hubble Space Telescope /
Carolyn Collins Petersen and John C. Brandt. – 2nd ed.
        p   cm.
    Includes bibliographical references and index.
    ISBN 0 521 59291 7
    1. Hubble Space Telescope (Spacecraft)   2. Space astronomy.
I. Brandt, John C.   II. Title.
QB500.268.P48   1998
522'.2919–dc21   98-21341   CIP

ISBN 0 521 59291 7 hardback

To the people who overcame tremendous political, financial, and technical problems to make the Hubble Space Telescope into a functioning observatory producing unique contributions to science on a routine basis.

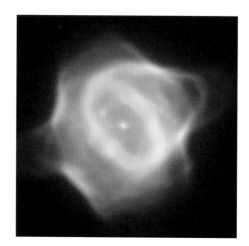

Contents

Preface

Astronomy is a remarkable business. Those of us who are privileged to work as astronomers (in whatever capacity we can contribute) marvel at the luck we have in being able to have so much *fun* doing our jobs! Through much of the 1990s, many astronomers have extended their views of the universe (and many of us have had fun doing so) using the Hubble Space Telescope. It has not always been an easy task. No one who has read a newspaper in the past few years can remain ignorant of the problems the HST has had. But, this mechanism — this complex collection of instruments and mirrors — has succeeded in opening the universe to us in ways we had never dreamed possible.

Our original goal in publishing *Hubble Vision* back in 1995 was to tell everybody that the telescope worked! We had each seen enough of the behind-the-scenes maneuvering, problem-solving, and triumphs to know that HST's story was one that needed telling — and not just the political, sometimes pejorative, tales that humans (being humans) love to tell and hear. There was (and is) good science being done, and that was the story we wanted to relate.

In the years since *Hubble Vision* was first published, the Hubble Space Telescope has brought us amazing sights from across the light years. Happily, this has made our job as science writers much more fun. Now we can focus on the joy of discovery that HST has awakened in the scientists who use it and the nonscientists who marvel at the images it returns to us.

The universe is a constantly changing place. HST is part of a long line of orbiting observatories – including International Ultraviolet Explorer, Hipparcos, Yohkoh, Compton Gamma Ray Observatory, Cosmic Background Explorer – that, along with an excellent collection of ground-based facilities, have carried out programs to study and map the growth and evolution of the cosmos. Of course, we cannot and do not ignore the problems that the telescope has faced. As we have heard from a number of scientists who use HST, it is a working observatory – and it faces challenges that no ground-based facility will ever experience.

Still, it is easy to be judgmental when discussing Hubble – after all, the telescope still operates in the full glare of publicity. In the early days, every media story about HST took pains to remind us about the mirror problems and the tremendous cost of the program. No other facility faced this sort of scrutiny, and no other observatory started its public life with such a spectacularly disastrous birth.

Now, having just said that we want to tell a science story, it would be shortsighted of us to ignore the political considerations completely. So, we have included a 'history of HST' chapter to help place the telescope in its proper sociological perspective. (For those readers who delight in reading about the politics and management of large programs, we highly recommend Robert Smith's detailed book, *The Space Telescope* (Cambridge University Press, 1993).) That leaves the rest of the book for the 'science goodies'. After all, that is why HST is up there – to get the goodies.

HST's scientific odyssey takes us well beyond our home neighborhood of planets to the stars and galaxies. The telescope's intriguing accomplishments – ranging from the discovery of more moons around Saturn, the aftermath of Comet Shoemaker–Levy 9's date with Jovian destiny, the detection of black holes at the hearts of distant galaxies, to the discovery of the most distant galaxies, and the ongoing refinement of the Hubble Constant – are the real story here. The good news about HST's journey is that there is a great deal of new science to discuss. The bad news is – like travelers who return from faraway places with lots of pictures and postcards – we cannot possibly talk about all of it in one sitting. In creating this 'Son of *Hubble*

Vision', our pile of postcards is a lot higher than it used to be. It makes our job a whole lot more interesting! We have chosen to present a survey of HST's observations to illustrate the scientific side of the Hubble Space Telescope's odyssey, and along the way we have tried to present the data in the larger context of just how the science of astronomy works.

We conclude our tale of HST's further adventures with a look at future plans for HST. To keep from turning this into a textbook, we have included an extended glossary at the end of the book that explains many of the concepts used in our discussions. When we talk about actual HST discoveries, we let some of the researchers who are the prime recipients of HST's data largesse talk about their work. Theirs are the voices not always heard in the media, yet they are the people who are using Hubble's data to 'push the envelope' of astronomical discovery.

Carolyn Collins Petersen
John C. Brandt

Acknowledgments

As usual, when authors take on a task like this one, a large number of individuals and institutions are called on to contribute their viewpoints, tolerate questions, and put up with the stresses that accompany any creative effort. Of course, our spouses come to mind first. Mark Collins Petersen and Dorothy Bell Brandt endured inevitable hassles and supported us both during this project. To them we owe our love and thanks.

Dozens of scientists, engineers, program managers, observers, and research personnel involved with HST took time to talk with us about their work and their experience using this most complex telescope. Our 'conceptual gestalt' of HST was built up through the conversations we had with these and many, many others both inside and outside the project.

Mike A'Hearn, Ron Allen, Claude Arpigny, Reta Beebe, Fritz Benedict, Bob Bless, Al Boggess, Alex Brown, Bob Brown, Margaret Burbidge, Jason Cardelli, Ken Carpenter, Bob Chapman, Art Code, Peter Conti, David Crisp, Cindy Cunningham, Roger Doxsey, Alan Dressler, Dennis Ebbets, Steve Edberg, Sandra Faber, Michel Festou, Otto Franz, Riccardo Giacconi, Heidi Hammel, Rich Harms, William Hathaway, Suzanne Hawley, Sara R. Heap, Jeff Hoffman, Bill Jeffreys, James Kaler, Peter Kandefer, Bill Keathley, Rob Landis, Tod Lauer, David Leckrone, Steve Lee, David Levy, Jeffrey Linsky, Duccio Macchetto, Richard McCray, Steve Maran, Bruce Margon, Georges Meylan, Heather Morrison, Claude Nicollier, Robert O'Dell, Jean Olivier, Cora Randall, Nancy Roman, Blair Savage, Jim Secosky, Carolyn Shoemaker, Gene Shoemaker, Sue Simken, Martin Snow, Ted Snow, Lyman Spitzer, Peter Stockman, Alex Storrs, Rodger Thompson, Larry Trafton, Bill van Altena, Hal Weaver, Ed Weiler, Jim Westphal, Ray Weymann, Robert Williams, Bruce Woodgate, and Ben Zellner took time out from busy schedules to share their vision of what it means to work with HST and to explain HST's place in astronomy history. We appreciate their gracious cooperation, and we apologize to anyone we interviewed who has been left off the list.

Many other scientists shared their science results and comments in the form of papers, press releases, and hallway conversations, and we found these materials to be invaluable. We also thank the many scientists who supplied us with illustrations, some custom-made just for this book. Special thanks go to Ray Villard and his staff at Space Telescope Science Institute's Office of Public Outreach for their extraordinary and enthusiastic cooperation.

Finally, we thank Mark C. Petersen for extensive manuscript review and suggestions, Tim Kuzniar for custom artwork, and Darien Gould for interview transcriptions. We are particularly grateful to the Cambridge University Press staff, in particular Adam Black, Irene Pizzie, and Simon Mitton. Astronomers Ken Carpenter, Sandra Faber, Carl Hansen, James Kaler, David Leckrone, and Steve Shore all offered superb assistance in reading parts of the manuscript and in bringing mistakes and potential problems to our attention. In a project this complex, a few errors are bound to creep in, and any remaining are, of course, our responsibility.

Hubble Space Telescope and the universe

Stars scribble in our eyes the frosty sagas, the gleaming cantos of unvanquished space.

Hart Crane

Light from distant places has made the journey to earth, and it falls on these new eyes of ours, the telescopes, the all-frequency time machines ... with these new eyes, we can make voyages as dramatic as those of the early navigators, out to the limits of the known universe.

Michael Rowan-Robinson

Figure 1.1. HST's view of Mars, taken on May 20,1997, during early summer in the Red Planet's northern hemisphere. *(Philip James, University of Toledo; Todd Clancy, Space Science Institute; Steven Lee, University of Colorado; NASA)*

Figure 1.2. Starbirth at the heart of Centaurus-A (NGC 5128). This galaxy is only 10 million light years away from us. The Near-Infrared Camera and Multi-Object Spectrometer peered through the haze of dust and gas surrounding a massive black hole at the center of the galaxy. This black hole is cannibalizing a smaller galaxy that collided with Centaurus-A. *(Ethan Schrier, Space Telescope Institute; NASA)*

Figure 1.3. A giant starbirth region in the neighboring galaxy M33. At the heart of this nebula (NGC 604) are more than 200 hot young massive stars. Their radiation heats the nebular gases, and highlights the nebula's shape. The image was taken with HST's Wide Field and Planetary Camera-2. *(Hui Yang, University of Illinois; NASA)*

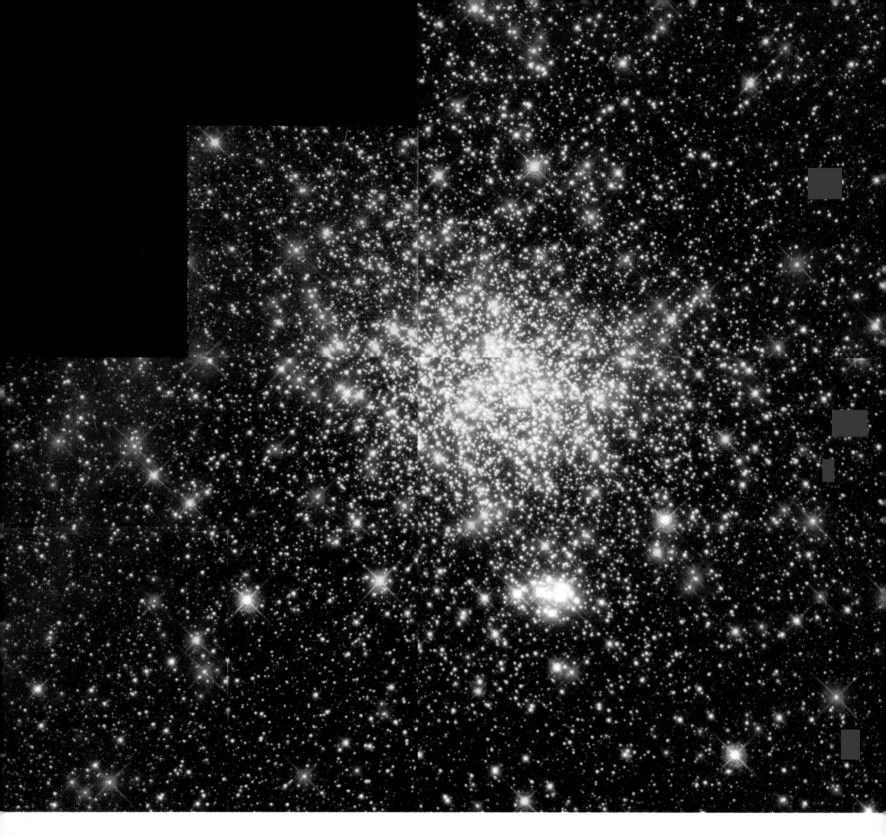

Figure 1.4. Two star clusters in the Large Magellanic Cloud as seen by the Wide Field and Planetary Camera-2. In the 130 light-year-wide view we see nearly 10 000 stars, many of which are part of the cluster NGC 1850. *(R. Gilmozzi, STScI/ESA; NASA)*

Figure 1.5. (Opposite) A pair of planetary nebulae imaged by Wide Field and Planetary Camera-2. **(top)** The 'butterfly' shape of M2-9 belies a pair of jets moving material from the center of the nebula at more than 300 kilometers per second. *(Bruce Balick, University of Washington; Vincent Icke, Leiden University (Netherlands); Garrelt Mellema, Stockholm University; NASA)*

(bottom) Ten thousand years of mass loss are shown in this image of CRL 2688, also called the 'Egg Nebula'. The arcs are shells of dense matter blown off the dying star. The bright 'searchlight' beams indicate the presence of holes in the clouds ejected by the star over a period of 10 000 years. *(R. Sahai, John Trauger, NASA-Jet Propulsion Laboratory; NASA)*

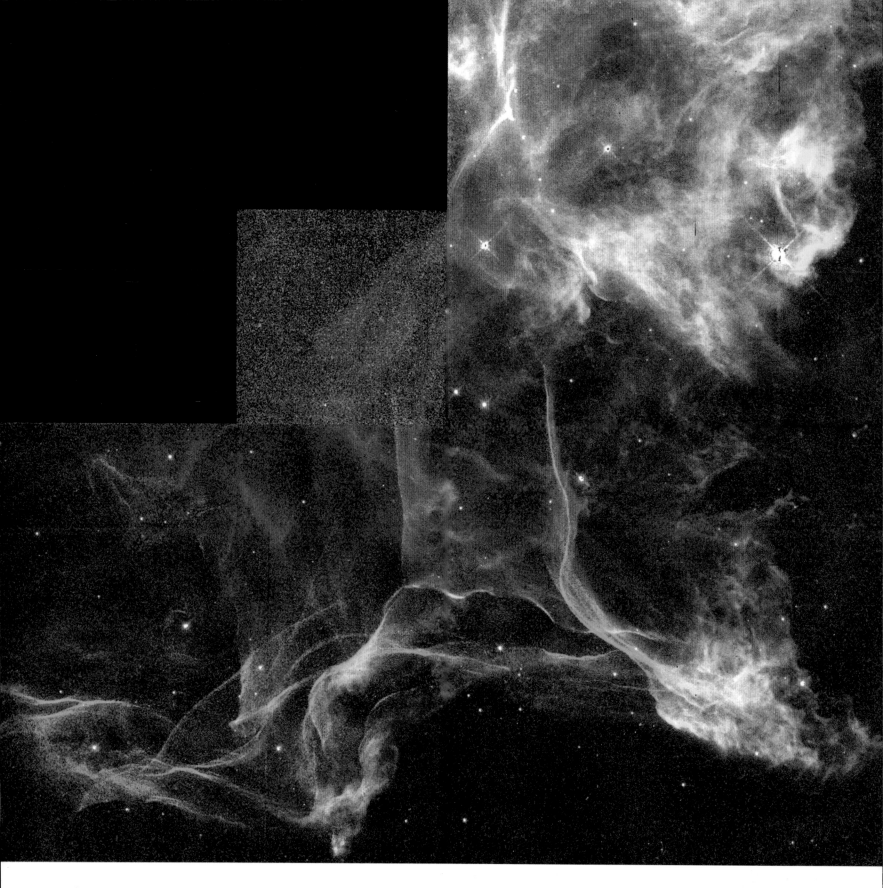

Figure 1.6. A small portion of the supernova remnant known as the 'Cygnus Loop', imaged by the Wide Field and Planetary Camera-2. Here, the supernova shock wave has collided with a dense interstellar gas, heating the gas and causing it to glow. *(Jeff Hester, Arizona State University; NASA)*

Figure 1.7. The aftermath of a galaxy collision, as seen by the Wide Field and Planetary Camera-2. The cores of the twin galaxies show up here as orange blobs. The sweeping spiral pattern is traced by bright blue star clusters whose birth was triggered by the collision. *(Brad Whitmore, STScI; NASA)*

Figure 1.8. A field of distant galaxies, with NGC 4881 at the center. These galaxies, which lie at the edge of the Coma Cluster (and beyond) may lie more than 100 megaparsecs away. *(William A. Baum, University of Washington, the HST WF/PC Team; NASA)*

Stargazing is a ritual as old as human culture. On a clear, dark night we lift our gaze to the skies, and suddenly our souls are touched by the splendor of the stars. We are the descendants of countless generations of astronomers, and as it did for our ancestors, the universe dangles tantalizing mysteries before our eyes. It dares us to come out and explore. Then, preposterously, it challenges us with seemingly impassable distances between Earth and anything but the closest planets. Yet, distances and light continue to fascinate us. We see something in the sky and ask, What is it? How far away is it? How fast is it traveling through space? How old is the universe? How big is it? Will it ever end?

Astronomy concerns itself with answering these questions and many others. It would be exciting if astronomers could explore the cosmos from the bridge of some spacefaring vessel like Star Trek's *Enterprise* – whipping between planets and star systems at warp speed and examining interstellar mysteries with long-range scanners and advanced probes. Unfortunately, experiencing the evolution of stars and galaxies on any short-term scale is impossible for us because we are a short-lived species. However, that has not discouraged stargazers from trying to make sense of the cosmos and exploring it with some of the most fantastic tools ever invented.

Hubble Space Telescope is one such tool, and it took its place among the world's premier observatories in 1990, amid great fanfare and scientific excitement. Hardly anyone interested in astronomy can fail to know that HST went through a rough 'shakedown cruise' period when problems were discovered with the telescope. Most publicized was the discovery of spherical aberration in HST's main mirror. Essentially, it was ground incorrectly. Instead of reflecting most of the light from an object into a well-focused 'core' of light, the mirror beamed a small amount into the core and spread the rest out into a diffuse halo around the core. This, in turn, cut down on the light going to the science instruments which depended on tightly focused light in order to turn out good observations. The spherical aberration, coupled with the breakdown of several other components in the telescope, created problems that were severe enough that the National Aeronautics and Space Adminstration mounted a complex mission in 1993 to correct them. The result of that mission, and another servicing mission in 1997 which introduced two new instruments, is an orbiting observatory that works every day, peering at objects as close by as our own Solar System and as distant as the observable limits of the universe! (For readers interested in HST's early years, we have included a history of the telescope and its supporting missions in Chapter 6.)

Many years before Hubble Space Telescope's odyssey began, a father and daughter stood outside on a balmy autumn night, looking up at the sky. All day the girl had heard the adults in the house talking about a strange new thing in the sky – something called 'Sputnik'. She begged her father to take her out and show 'Sputnik' to her. Finally, when it was dark enough, the two went out, and he tried to tell his daughter that Sputnik was up there, but that it would be hard to see. All his daughter could see were stars, even though she was not sure what they were. What *he* saw was a little girl who wanted something she could not understand. But they stayed out there for a while that night, looking for a bright, shining point of light, moving across the sky – Earth's first artificial satellite. They never saw it, but it was her introduction to the night sky, and it expanded her universe for ever.

More than 40 years have passed since Sputnik raced skyward, and humans have been sending instruments and animals and people into space on a regular basis ever since. Like that of the little girl, our understanding of the cosmos changes with every space mission. Not only is the universe more complex than we ever dreamed, but the instruments we use to study it have evolved from a simple telescope into intricately interlocked systems of optics and electronics – multifrequency detectors that open for us a window into space and time.

The complexity of the cosmos is reflected throughout the universe, from the microscopic world of the atom to the macrocosmic domain of the galaxies. To fully appreciate that breathtaking range, we are compelled to transform it into units we can understand. The simplest are, of course, the ones by which we measure our daily lives. We wake up at home, we leave to go to work or school (usually a few kilometers away), we travel from one country to another (perhaps thousands of kilometers away). On a slightly larger scale is the Solar System, stretching across thousands and millions of kilometers. From there, we cast inquisitive eyes from our little planetary niche across hundreds of millions of kilometers to the stars in our own galaxy. Beyond the Milky Way, other galaxies wheel through the cosmos in clusters, millions and billions of kilometers away. Just at the faintest edge of our astronomical eyesight, we find evidence of the birth of the universe (many, many billions of kilometers away). Of course, what we are really searching for is an understanding of the processes that constantly change the cosmos.

We see the seeds of change, of evolution, every day. An earthquake destroys a city, rains wash away homes, and hurricanes devastate coastal areas. Great sprawling cities and highways, dams and bridges change the face of the Earth as well. On the surface of a planet like

Figure 1.9. The size and scale of cosmic systems. Astronauts servicing the HST make up a scene that is merely a point within the cube containing the Earth, which is itself a point in the cube containing the Solar System. In turn, our planetary system is just a point in the stellar neighborhood, which is itself a point within an arm of our Milky Way Galaxy. The Milky Way is a point in the galactic neighborhood, which, finally, is just a point in the entire universe. *(Dana Berry, STScI)*

Mars, we find the evidence of a watery past, completely different from what exists today. In one part of the sky, an immense cloud of gas and dust turns out to be the birthplace of stars, while elsewhere an ominous-looking cloud is all that is left of a supergiant star that died in a violent cataclysm. In the beauty of a nearby galaxy, we see the seeds of its birth. How do we see all this? It all comes down to using and understanding the properties of something we take for granted – light.

Astronomy and light

Light is the Rosetta Stone of astronomy – the guide to understanding the complexity of the universe. Locked within the light that surrounds us is an incredible amount of information about the objects that radiate and reflect it. We can learn a great deal of information about a light source – ranging from its temperature and chemical makeup, to its speed and direction, as it travels through the universe. The history of astronomy itself can be characterized as the history of humans learning how to decode the mysteries of light. Once we knew light was the key to the universe, we developed a vast array of devices to capture and analyze it in all its forms.

The light most familiar to us is what we see with our own eyes. This is called 'visible light'. It is part of a larger array that astronomers call 'light' which runs from the shortest wavelengths of about 10^{-8} meter to the longest wavelengths of 3 millimeters. Astronomers' 'light' is, in turn, part of a larger array called the *electromagnetic spectrum* (commonly written as EMS). Astronomers also refer to very small wavelengths of light by another unit – the angstrom, which is equivalent to 10^{-10} meter. Thus, the shortest wavelengths for light would be 100 angstroms. The EMS encompasses everything from gamma rays, x-rays, and ultraviolet light through visible wavelengths, infrared, and radio waves. To put it into perspective, imagine if

Figure 1.10. The electromagnetic spectrum (EMS). Here is the entire range of electromagnetic waves, relevant scales and the objects most likely to emit them. The Earth's atmospheric opacity, which determines what radiation reaches the surface, is shown as a silvery band. Visible light and radio waves reach the surface, while most other waves do not. *(Dana Berry, STScI; Greg T. Flynn, Sky Publishing Corporation)*

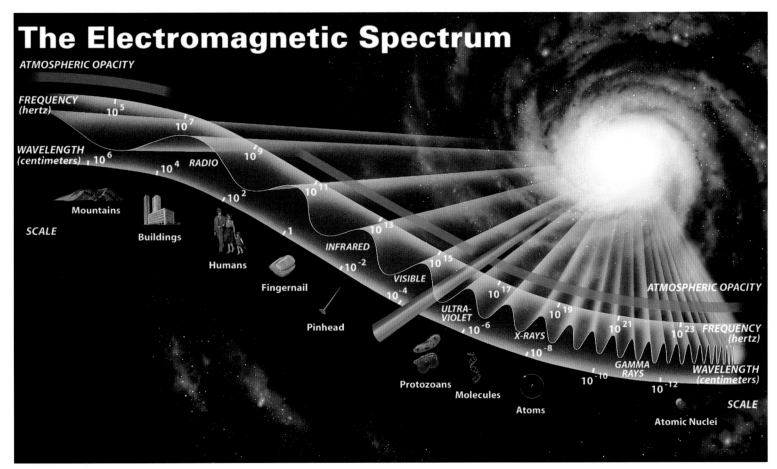

all the wavelengths of the electromagnetic spectrum – up to radio wavelengths of 100 meters – could be spread out across an area about the length of an American football field (91.4 meters). Visible light – the light we can see with our eyes – lies in a band between 4000 angstroms and 7000 angstroms. This band is 0.000 000 3 meters wide – much narrower than one of the blades of grass on that imaginary football field.

Light, however, is not totally defined as a collection of wavelengths. Certainly it can behave as a wave, but it can also act as a particle. Actually, it can act as both a wave and a particle at the same time; the only difference lies in how we observe it. A particle of light is called a *photon*, which can be thought of as a packet of energy. Each photon has an energy value, given in units called *joules*. The photon also has an underlying wave nature – i.e. it exhibits the characteristics of its wavelength. Red light, for example, has a wavelength of about 6500 angstroms; a photon of that red light has about 3×10^{-19} joules of energy. The common light bulbs used in our homes radiate yellowish-white light, which consists of photons that are just a little more energetic than red light. When you turn on a 100-watt light bulb, it emits about 10^{20} photons every second and dissipates the rest of its energy as heat.

To form some idea of the duality of light, imagine going down to the beach on a warm summer's day. As you sunbathe, photons of visible light from the Sun hit your eyes and enable you to see your surroundings. You also feel infrared wavelengths as warmth from the Sun, and some ultraviolet radiation burns, or tans, your skin. The Sun also emits higher-energy ultraviolet and gamma radiation, but you do not experience it as you lie there on the sand because the Earth's atmosphere screens it out.

Now turn that experience to stargazing. You stagger outside with your observational accoutrements – telescope or binoculars, star charts, hot drinks, chocolate bars, blanket, radio or CD player – and start an evening of observation. You are very likely going to be observing objects that radiate in the visible wavelengths of light. If you watch the Moon or the planets, you are looking at reflected light because these objects do not generate their own light. Stars and galaxies provide you with radiated light because they do generate their own photons.

It is not much different for professional astronomers (and the many advanced amateurs out there!) who simply use bigger telescopes and more complex instruments to study the universe. But, instead of limiting their view to visible wavelengths, these astronomers try to study every wavelength of light, using both ground-based and orbital instruments.

Astronomy: the observational science

The principle behind the array of astronomical instruments in use today is fairly simple: gather as much light as you can from objects in the universe, and analyze it. What sort of light you gather depends on what you are using to gather it. There is no ideal light detector, no instrument that senses all wavelengths of light perfectly. So astronomers use different detectors to study light of various wavelengths.

Telescopes are the most familiar astronomical tools around. At its heart, a telescope is simply a device that gathers electromagnetic radiation – usually in the form of ultraviolet, visible, infrared, or radio wavelengths. For 'optical' devices, a mirror at the heart of the instrument focuses and reflects light to other sensors, such as film cameras and electronic recorders called CCDs. A CCD (charge-coupled device) is a specialized camera used on a wide variety of telescopes. Within this camera is an extremely sensitive chip that gathers light across the 'picture elements', or pixels, of its surface. The image is 'read out' from the chip and transferred to an electronic storage medium. Like other cameras, the CCD captures an astronomical event at a single fixed point in time. From the images produced by cameras, astronomers can learn about such things as brightness of an object, its shape, location, and relationship in space to other nearby objects.

The bigger the telescope you build, the more light you can gather. However, some very practical considerations limit the size of the instrument and its placement. To begin with, any telescope mirror has limits on how big it can be. A large mirror can gather more light, giving its users a chance to look at dimmer and more distant objects. However, mirrors cannot be made infinitely large because the structures to support them would have to be infinitely large as well. The shape of a mirror also limits its ability to concentrate incoming light to a tight focus, and large mirrors bend and sag under their own weight.

Observatories function best when they are located away from sources of light, heat, and radio wave pollution. Ideally, observatories should also be at high elevations to minimize the effects of Earth's atmosphere on the incoming light. Many of the world's facilities are located on mountain tops (such as the collection of telescopes at Mauna Kea, Hawaii; Cerro Tololo, Chile; and Pic du Midi, France). Others, like the Very Large Array near Socorro, New Mexico, are located well away from cities in wilderness areas and deserts. The latest twist on 'getting away from it all' involves the use of space-based observatories. Spacecraft – both Earth-orbiting and so-called 'fly-by' missions – have carried a variety of instruments and sensors to study light of wavelengths that are filtered out by the Earth's atmosphere.

Basically, astronomical observations can be divided into three categories: *imaging, photometry*, and *spectroscopy*. In imaging, the light is recorded on film or by an electronic camera such as the CCD we mentioned earlier. We now know that some objects in the universe change rapidly, and scientists learn more if they can study an object or event as it evolves. Certainly, telescopes can image something many times over the course of an observation, or for a much longer time than the human eye can gaze at an object, but there are limits to just how many 'pictures' can be taken with ordinary film cameras, plates, and single CCDs. So, specialized cameras are used to capture an event as it happens over a period of time. This is the concept behind high *temporal* (time) resolution, and the best systems will give us many 'snapshots' of an event taken within fractions of a second.

Sometimes a telescope is pointed at a field crowded with objects, such as a star cluster. If we are studying just one star in the cluster, it helps if the instrument can separate one star from another. This is the concept of *spatial* resolution – the ability to distinguish objects in a crowded field from each other and to produce clearly defined images of them.

Photometry is the practice of measuring the intensity of light from an object, and it often depends on instruments that can record with high temporal resolution. A photometer can be thought of as a very sensitive light meter, similar to that used in flash photography. Photometers are used to determine fluctuations in light intensity from a variable star, for example. The determination of intensity of light from stars, and other objects in the universe, is extremely important because from those studies comes the determination of brightnesses, or magnitudes. Quite often an astronomer will refer to an 8th-magnitude star or a 5th-magnitude comet. What the number refers to is the relative brightness of the object compared to other objects. The brightest stars have the smallest numbers, while dimmer stars have larger magnitude numbers. The brightest star in the night-time sky is Sirius, with a visual magnitude of −1.5; Canopus is −0.7; and Betelgeuse is around 0.5. The dimmest stars we can see with the naked eye range around 5th or 6th magnitude.

Different photometers measure different wavelengths, so there are whole sets of instruments sensitive to infrared and ultraviolet light as well as to visible light. Photometry is a very systematic way of classifying celestial objects and of observing how their changes of light output affect the way they appear to us. Before it was replaced by COSTAR, HST's High Speed Photometer worked in this way, recording the brightnesses of objects at high temporal resolution.

Spectroscopy takes light and divides it into its component wavelengths. We are all familiar with some everyday types of spectroscopy: white light shining through a prism, and sunlight shining through raindrops to form a colorful rainbow, for example. These demonstrate very simply that the Sun's light (or light from any white-light source) radiates all the wavelengths in the spectrum.

Spectroscopy answers a variety of questions that might not be answered if we relied solely on undispersed light to study objects: what is the chemical composition of an interstellar gas

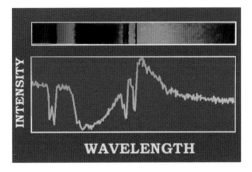

Figure 1.11. Much of the data gathered by HST are spectroscopic. This is what a spectrum looks like. The black lines in the color bar correspond to dips in the graph. The lines create patterns of absorption or emission, which tell astronomers which elements are present in the object being observed. Depending on the position of the lines, astronomers can also tell if the object is moving away from or toward the Earth, as well as its velocity relative to us. *(Dana Berry, STScI)*

cloud? the temperature of a star? the chemical makeup of a comet? the velocity of a gas jet at the center of a galaxy? The methods of spectroscopy allow us to answer these questions by studying the way in which stars and galaxies, comets and planets emit and absorb light.

The simplest spectrographic tool is the spectroscope, which breaks light down into very fine divisions by wavelength. A prism will work, but most modern astronomical spectrographs – instruments that collect light, break it down and record the results on film or as computer data – use diffraction gratings. These are mirrors scored with very thin lines. As the light shines across the grating, it is broken up into very, very fine wavelength divisions. You can begin to see how a diffraction grating works by looking at a compact disc in sunlight. A very fine, continuous rainbow of colors appears.

Spectroscopy is a very powerful tool in chemistry, where the identifying characteristics of an element can be determined with great precision. It is a fairly simple process in the lab – all you do is apply heat to an element and study the light given off as the element burns. Each element has a very distinctive 'fingerprint', or spectrum. Generally, a spectrum looks like a smooth continuum of color, broken by very bright or dark lines. The story that spectroscopy reveals about an object is found in these lines. The concept of *spectral* resolution applies in spectroscopy – the ability to cleanly separate adjacent features in a spectrum.

The basic rules of spectroscopy were written in the 19th century by a German chemist named Gustav Kirchhoff. His Laws of Spectral Analysis describe what sort of spectra appear as elements are burned. Kirchhoff's first law states that a hot, high-density gas or an incandescent solid body will radiate across a continuous spectrum. We will see light of all wavelengths in its spectrum. The second law states that a hot, low-density gas will produce what is called an 'emission-line' spectrum, i.e., elements that are abundant in the object will appear as very bright lines in its spectrum. (Parts of the Orion Nebula, for example, shine very brightly in emission spectra.) Kirchhoff's third law states that when a source of continuous radiation, such as a star, is viewed through a cooler, low-density gas, an absorption-line spectrum will be produced. The study of absorption spectra is a particularly ingenious way of determining what lies between us and a star. The spectrum of that star will show dropouts where certain wavelengths of light have been absorbed by clouds of material in interstellar space. All an astronomer needs to do to make an identification is to compare that spectrum with laboratory spectra of elements suspected to exist in the star.

HST: the multi-frequency time machine

Like its Earth-bound counterparts, the Hubble Space Telescope makes its observations with mirrors and detectors. It operates in just the same way as other telescopes – by gathering light from objects for analysis by science instruments. Because HST's science instruments are sensitive to a wide range of light, it is often said it gives astronomers an 'extended, high-resolution view' of the universe. Since HST was designed to operate in orbit around the Earth, it has a few extra features that ground-based telescopes do not have – solar panels to generate electrical power for everything on board, and magnetometers to detect the Earth's magnetic field. However, it has other features that Earth-bound astronomers would recognize – its own dedicated computer, gyros and sensors to help position it and lock on to guide stars (similar in spirit to the setting circles and finders used to help position smaller telescopes), and a communication link to Earth (just like the telecommunications lines between facilities on the ground). Like most Earth-bound facilities, it can even cover its aperture when it needs to – just like capping the telescope tube or covering a mirror on the ground.

As with other Earth-bound facilities, HST has controllers, operators, and users. The Space Telescope Science Institute (STScI) is the main access center for the telescope. Housed on the Johns Hopkins University campus in Baltimore, Maryland, the Institute is charged with allocating time on HST, and working with the scientists who use the telescope. Data from HST are stored here, and all of the public outreach functions associated with such a high-profile

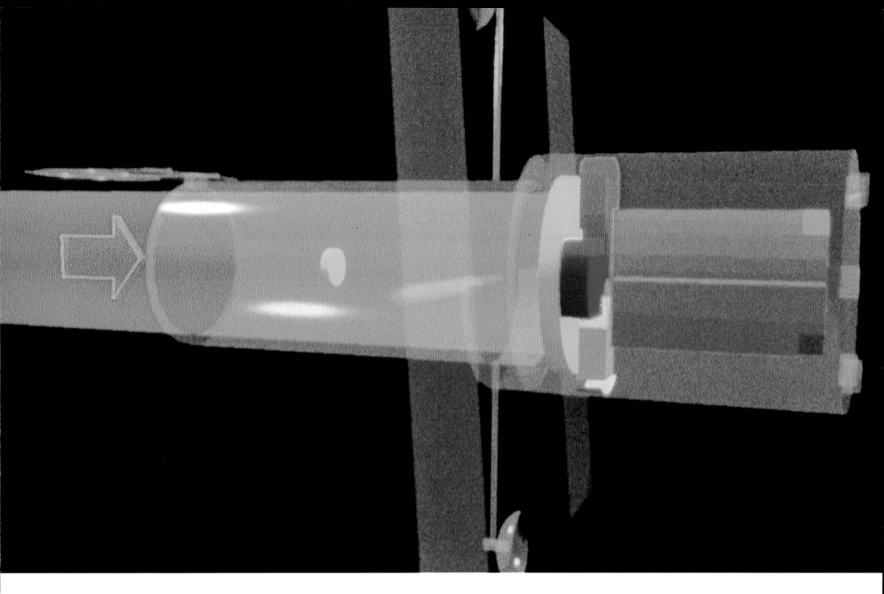

mission are performed at STScI. The Institute works in tandem with a control center at the nearby NASA Goddard Space Flight Center that has responsibility for maintaining orbital control of the spacecraft.

The heart of the telescope is the mirror system. HST's mirror is part of the Optical Telescope Assembly. It is a special Richey–Chrêtien version of a standard Cassegrain-type telescope. In a Cassegrain telescope, light from an object enters the telescope tube and bounces from a primary mirror to a secondary mirror, which sends the light back through a hole in the primary mirror and focuses it onto an imaginary surface, the focal plane. The HST system was built to provide light in the focal plane of the telescope that is close to the limit specified by the laws of physics – or, as the telescope scientists call it, the 'diffraction limit'.

The diffraction limit is a physical property of light that depends on the size of the primary mirror and the wavelength of the incoming light. As we mentioned earlier, when light hits the primary mirror on HST, it is supposed to be focused down to a small dot of encircled light energy on the focal plane. Theoretically, HST's mirror could concentrate around 85 per cent of the incoming light into this tiny circle. This is the diffraction limit of the mirror, and it is supposed to guarantee delivery of the highest possible quality image to the science instruments. As everyone found out in 1990, the mirror spread the light out. The corrective optics installed during the First Servicing Mission (in 1993) worked around the error by focusing 84 per cent of the light back into the circle. New instruments installed on HST such as the Wide Field and Planetary Camera-2, the Near-Infrared Camera and Multi-Object Spectrometer, and the Space Telescope Imaging Spectrograph, have internal optics to correct for the spherical aberration.

Figure 1.12. The light path through HST. The arrow indicates the direction of the incoming light, which is reflected by the primary mirror back toward the secondary mirror (small light-blue disc). The secondary sends the light through a hole in the primary mirror to the science instruments (indicated by red, yellow, green, and orange boxes).

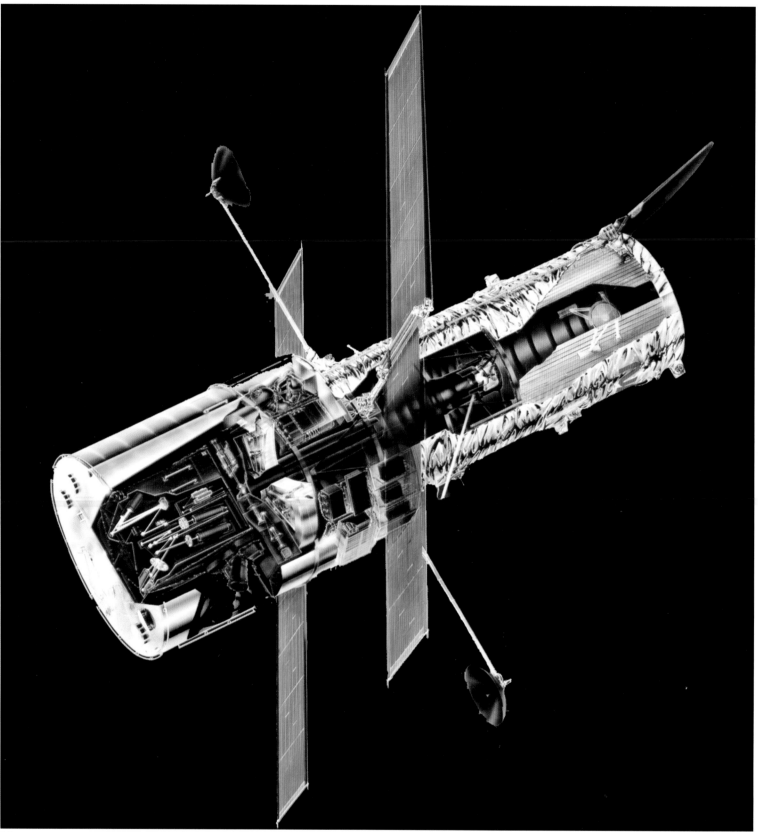

(a)

Figure 1.13. (a) A cutaway schematic of the Hubble Space Telescope. **(b)** The key to the various parts of the telescope. *(NASA)*

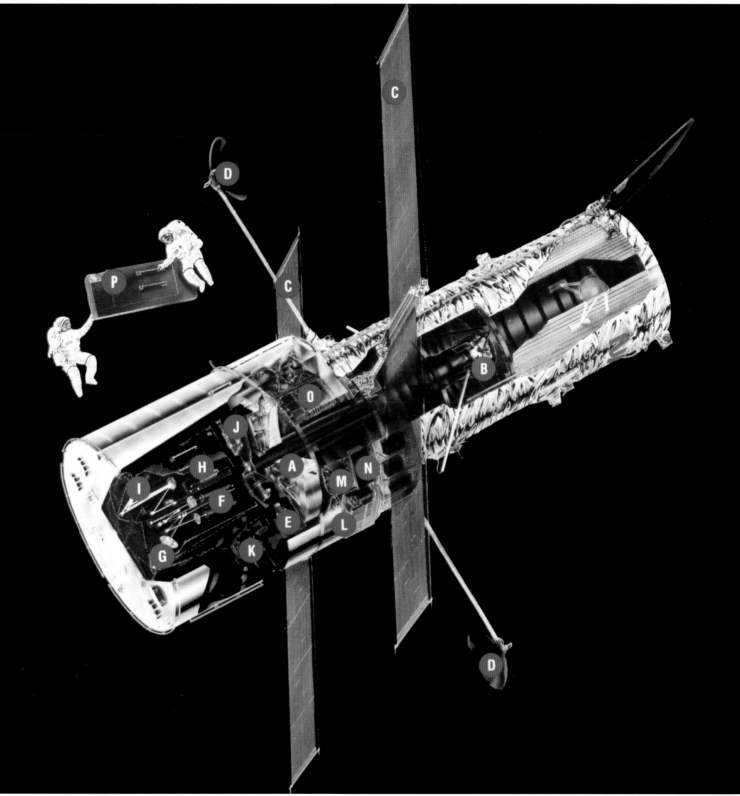

(b)

A. Primary mirror
B. Secondary mirror
C. Solar Arrays
D. Communications Antennas
E. Wide Field/Planetary Camera
F. ESA Faint Object Camera
G. Faint Object Spectrograph/Near-Infrared Camera

and Multi-Object Spectrometer
H. High Speed Photometer/Corrective Optics Space Telescope Axial Replacement (COSTAR)
I. Goddard High Resolution Spectrograph/Space Telescope Imaging Spectrograph
J. Fine Guidance Sensor (1 of 3)
K. Fixed Head Star Trackers (3)

L. Nickel Hydrogen Batteries (3 of 6)
M. Spacecraft Computer
N. Data Management Unit
O. Science Computer
P. Modular Replacement Instrument

The HST science instruments

HST was designed to carry five science instruments. Since its 1990 launch, the telescope has had eight different instruments installed at various times. The best known is the Wide Field and Planetary Camera (WF/PC). This is the instrument that gives us HST's visual take on the universe, and two have been used on HST – WF/PC-1 and WF/PC-2. The camera is a radial instrument, which means that it is mounted behind the main mirror, perpendicular to the long axis of the telescope. The other instruments are mounted parallel to the long axis of the spacecraft behind the mirror, and are called axial instruments. The Faint Object Camera (FOC) does visible imaging, but is sensitive to ultraviolet light as well. The Faint Object Spectrograph (FOS) operates over a wide wavelength range, and the Goddard High Resolution Spectrograph (GHRS) is a purely ultraviolet-sensitive instrument. The FOS and GHRS were replaced during the Second Servicing Mission (1997) by the Near-Infrared Camera and Multi-Object Spectrometer (NICMOS), and the Space Telescope Imaging Spectrograph (STIS), respectively. The High Speed Photometer (HSP) was HST's 'light meter'. Although not built as science instruments, the Fine Guidance Sensors (FGSs) serve as star trackers, and also perform astrometry – the science of accurately measuring stellar positions. During the First Servicing Mission (1993), WF/PC-2 replaced WF/PC-1, and the High Speed Photometer was removed to make room for the COSTAR system.

The Wide Field and Planetary Cameras

All of HST's instruments overlap each other to some degree in function and wavelength range, allowing scientists to collect the maximum data relating to an object. During the first three years on orbit, WF/PC-1 was the most-used instrument on the spacecraft, and WF/PC-2, with its own built-in spherical aberration correction system, is continuing that tradition.

Figure 1.14. A schematic of the field of view seen by the WF/PC-2. The dashed line at upper right indicates the field of view of the 'WF' portion of the WF/PC-1. *(T. Kuzniar, courtesy of Loch Ness Productions)*

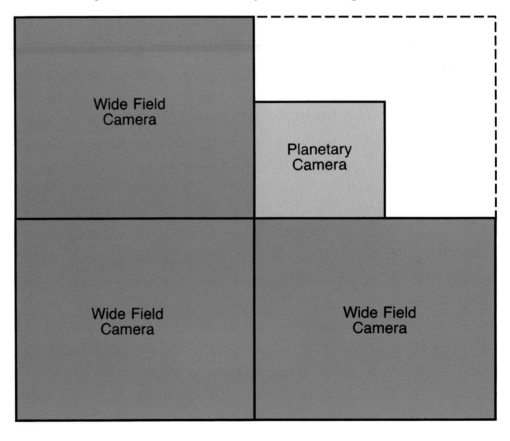

Figure 1.15. A sample WF/PC-1 image. This is the central region of the remote cluster of galaxies CL 0939+4713. This cluster is discussed further in Chapter 4. *(Alan Dressler, Carnegie Institution; NASA)*

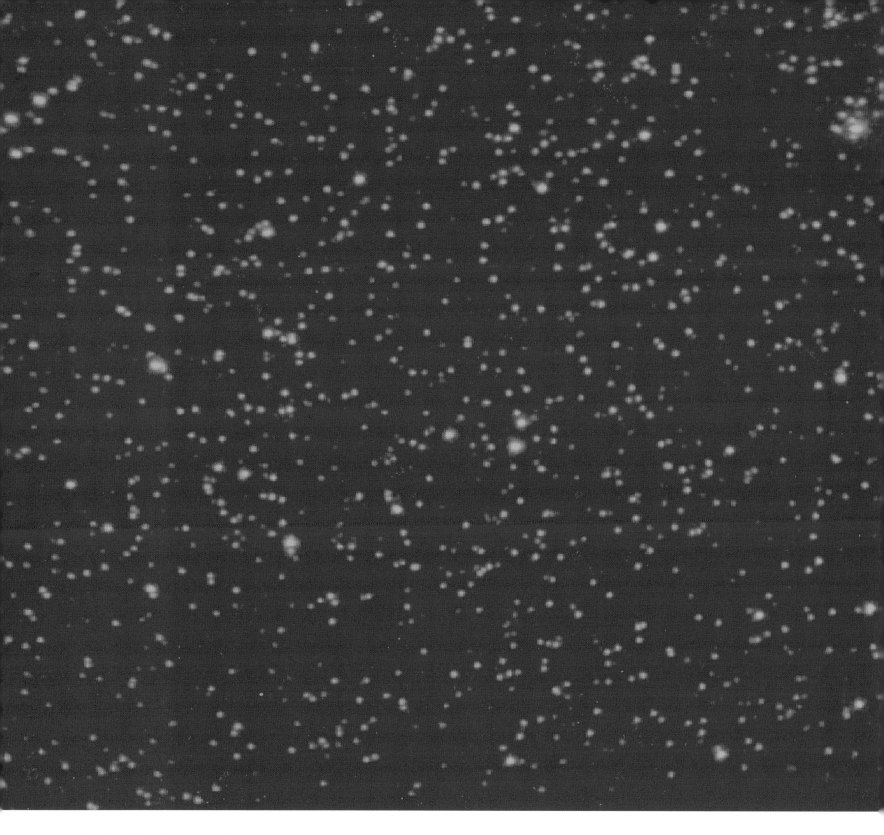

Figure 1.16. A sample FOC image (post-COSTAR) of the core of the globular cluster 47 Tucanae. The image is 14 arcseconds on a side and shows individual star images. The brightness of the stars has been measured accurately, and some of them are considered to be white dwarfs. *(R. Jedrzejewski, STScI; NASA; ESA)*

Both Wide Field and Planetary Cameras were built by the NASA Jet Propulsion Laboratory in Pasadena, California. WF/PC-1 was designed under the direction of the Investigation Definition Team headed by California Institute of Technology professor James Westphal, who later won a MacArthur Genius Grant for his work on the camera.

Because NASA expected WF/PC-1 to deliver images and data of wide appeal to the public, its replacement in case of failure was a high priority. While WF/PC-1 was still under construction, NASA provided funds for a 'clone,' which became WF/PC-2. The discovery of spherical aberration early in WF/PC-2's fabrication process allowed that instrument team, headed by Jet Propulsion Laboratory scientist John Trauger, to install corrective optics inside the instrument. For his work on behalf of the team that designed the second camera, Trauger was awarded the 1997 Masursky Award by the Division of Planetary Sciences of the American Astronomical Society for advances in and contributions to planetary science.

To accomplish the kind of work that everyone wanted to do with the WF/PCs, the teams designed their instruments to operate in two modes – wide-field mode or the higher-spatial-resolution planetary camera mode. Theoretically both WF/PC-1 and WF/PC-2 are sensitive to wavelengths of light ranging from about 1200 angstroms (in the ultraviolet) to just under 11 000 angstroms (in the infrared). This range also encompasses visual wavelengths. The WF/PCs have been responsible for some of the most spectacular images released from HST. The best spatial resolution of the WF/PC-2 is 0.053 arcsecond.

WF/PC-1 used two sets of four CCDs as its main light receptors, giving a full picture with an array of 1600×1600 pixels across the four CCDs. The replacement instrument, WF/PC-2, uses four CCDs at two different magnifications. There are three wide-field camera fields on the new instrument and one planetary camera field, which is why some images from the instrument appear to be 'chevron-shaped'. Nonetheless, WF/PC-2 allows its users to see objects as faint as 28th magnitude during long-term exposures.

The Faint Object Camera

The Faint Object Camera (FOC) was built for HST by Dornier Corporation and was funded by the European Space Agency. The Investigation Definition Team responsible for its design was chaired by H.C. van de Hulst of the Leiden Observatory in the Netherlands. The FOC is part of the European contribution to the spacecraft, and its operations are overseen at STScI by Duccio Macchetto, who has been with the project for many years, and acts as Principal Investigator for the instrument.

The FOC is designed to provide high-resolution images of small fields, and is sensitive to light in a range of 1150–6500 angstroms. It has two complete detector systems with image intensifiers to gather as much light as possible from dim, distant objects. Its maximum field of view is a square measuring about 14 arcseconds on each side, providing its highest spatial resolution of 0.042 arcsecond. The FOC's image is produced on a phosphor screen, and that screen is scanned by a specialized television camera. If the FOC looks at anything brighter than 21st magnitude, filters must be used to dim the light so that the detectors do not become saturated.

As with other instruments on the HST, the FOC suffered from the effects of spherical aberration. Currently, it is the only instrument on the telescope which requires refocused light from COSTAR.

The Goddard High Resolution Spectrograph

Because spectroscopy is such an integral part of modern astronomy, four spectrographs have been used on HST. During the first six years of the mission, the Goddard High Resolution Spectrograph and the Faint Object Spectrograph supplied users with otherwise unobtainable ultraviolet views of the universe.

Figure 1.17. A sample of data from the Goddard High Resolution Spectrograph. The spectrum of Betelgeuse (Alpha Orionis), shows the bright reddish star in the shoulder of Orion. The spectrum shows the bright and dark regions of the star's spectrum (below), and how the same information appears in the graphical form used by astronomers. Some of the elements in the star's atmosphere by GHRS are Fe for iron and Mn for manganese. *(Martin Snow, University of Colorado and GHRS Science Team)*

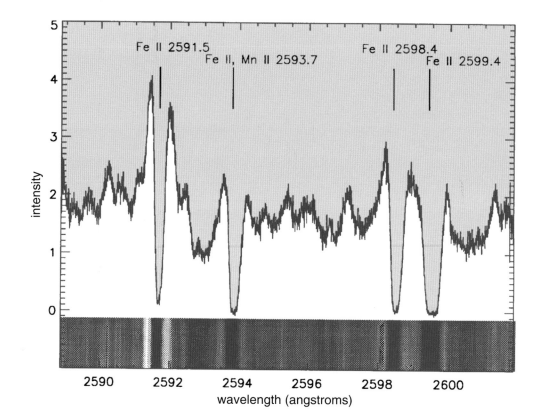

The Goddard High Resolution Spectrograph (GHRS) was built by Ball Aerospace Division, Boulder, Colorado, under the direction of an Investigation Definition Team headed by John C. Brandt (who was at Goddard Space Flight Center at the time). When Brandt moved to the University of Colorado in 1987, Goddard scientist Sara Heap was named as Co-Principal Investigator.

The GHRS consisted of an optical system, support electronics, and a structural system. The way it worked was simple: ultraviolet light entered the spectrograph through one of two apertures and was sent to a rotating carousel by a collimator mirror. The carousel had gratings that spread the ultraviolet light out. Depending on the gratings chosen, observers selected certain wavelengths of ultraviolet light for specialized study. The GHRS was sensitive to light between 1150 and 3200 angstroms. In medium resolution mode, it traced features 0.1 angstrom wide, and in high-resolution mode it traced features 0.02 angstrom wide.

The Faint Object Spectrograph

The Faint Object Spectrograph (FOS) studied fainter objects than could be detected by the GHRS. The FOS was built by Martin Marietta Astronautics Group of Denver, Colorado, under the direction of an Investigation Definition Team headed by Richard Harms, who was initially at the University of California-San Diego. The FOS was designed to be sensitive in the wavelength range of 1100 to 8000 angstroms, more extensive than the GHRS range. There was a tradeoff involved with this wider wavelength range: the FOS studied objects at a lower spectral resolution of light than did the GHRS, but the GHRS looked 'in depth' at objects in a limited wavelength range of ultraviolet light and at higher spectral resolution.

The FOS operated in two resolution modes: low and high. In low-resolution mode, it imaged 26th-magnitude objects in 1 hour exposures and traced features 15–20 angstroms wide. In high-resolution mode, it achieved 22nd-magnitude viewing in a 1 hour exposure and traced features 3–4 angstroms wide.

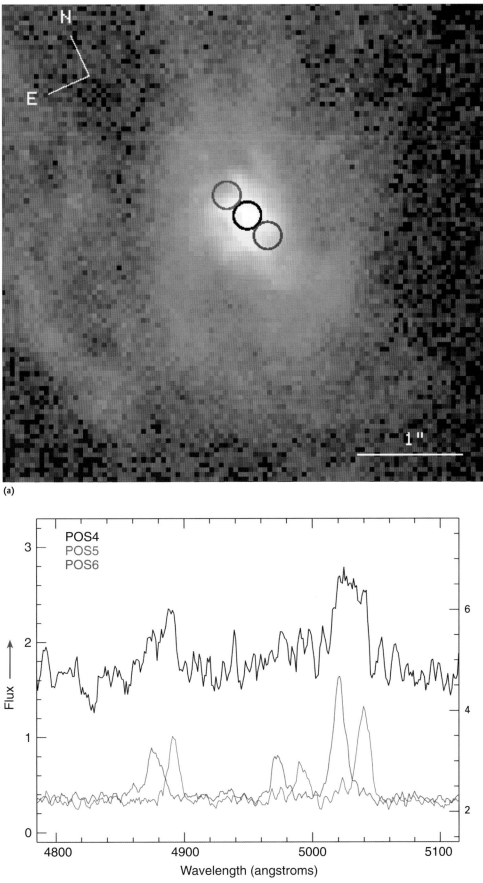

(a)

(b)

Figure 1.18. Sample FOS data: a spectrum of the center of active galaxy M87. The locations and size of the regions are color coded (red, blue, black) to the plot. The red spectrum area clearly shows redshifted light coming from a region that is moving away from us. The blue spectrum shows a blue shift, indicating light from material that is moving toward us. The black spectrum shows no systematic motion. The velocity of the material moving away from us is approximately 550 kilometers per second. This sort of evidence is important in establishing the existence of a massive black hole at the center of M87 (which we discuss in more detail in Chapter 4). *(R.J. Harms, RJH Scientific, Inc.)*

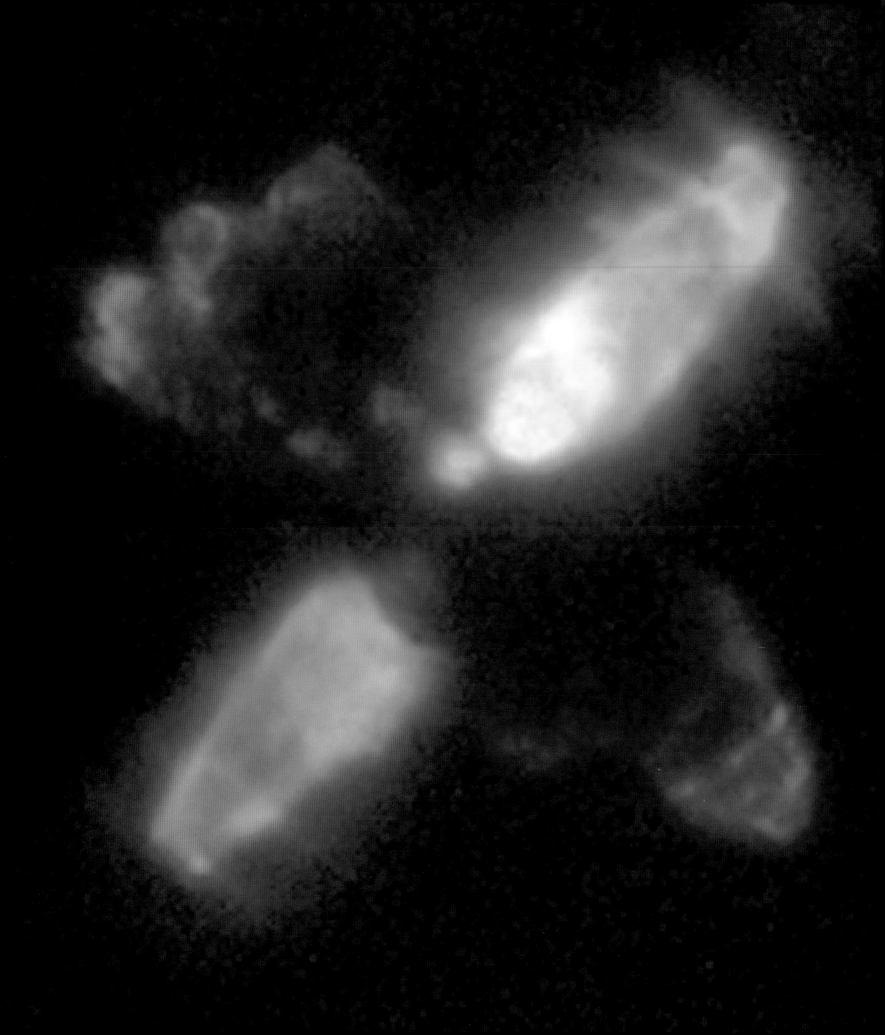

The Near-Infrared Camera and Multi-Object Spectrometer

Hubble Space Telescope came into its own as a 'multi-frequency' observatory with the installation of two more instruments in 1997. The Near-Infrared Camera and Multi-Object Spectrometer replaced the FOS, providing the telescope with its only infrared capability. Its spectral range runs from 8000 to 25 000 angstroms, and its largest field of view is 52 × 52 arcseconds. NICMOS uses a variety of filters, polarizers, and grisms to study the light from objects as faint as 25th magnitude at a resolution of 0.14 arcsecond.

NICMOS has three cameras to provide the highest possible resolution sampling of light in the infrared range of the electromagnetic spectrum. To achieve its goals, the instrument's detectors are cooled with solid nitrogen. It also contains special optics to correct for the effects of the main mirror's spherical aberration.

NICMOS was designed and built at Ball Aerospace in Boulder, Colorado, under the direction of University of Arizona astronomer Rodger Thompson.

Space Telescope Imaging Spectrograph

The Space Telescope Imaging Spectrograph (STIS) is a wide-band, multi-purpose imaging instrument, also built by Ball Aerospace, under the direction of Goddard Space Flight Center's Bruce Woodgate. It studies light in the wavelength range of 1150–11 000 angstroms, making STIS a functional replacement for both the GHRS and the FOS, although it physically replaced the GHRS. In reality, it provides an ideal way to study a range of objects from active galactic nuclei to comets.

STIS is a two-dimensional spectrograph. Light is passed through a long slit on the instrument, and STIS can then produce simultaneous spectra of specific points within a light source. The instrument's widest field of view is 50 × 50 arcseconds, and it can detect objects down to magnitude 28.5. Both NICMOS and STIS have coronagraphic capability, which means that they can block out unwanted light from bright sources when studying a dim, distant object of interest.

Figure 1.19. (Opposite) The Egg Nebula, seen through the infrared eyes of the Near-Infrared Camera and Multi-Object Spectrometer. *(Rodger Thompson, University of Arizona; NASA)*

Figure 1.20. A high-resolution STIS spectrum of a star (HD 72089) behind the Vela supernova remnant. The entire spectrum from 1196 to 1397 angstroms is presented. In this format, the full spectrum appears in strips with the lower wavelengths at the bottom and the higher wavelengths at the top. The dark strips near the bottom show heavy absorption by interstellar neutral hydrogen. The interval marked by the box is shown graphically with a wavelength. The absorption line at approximately 1277.3 angstroms is attributed to absorption by neutral carbon in the supernova remnant. *(Edward B. Jenkins, Princeton University Observatory; Martin Snow, University of Colorado; NASA)*

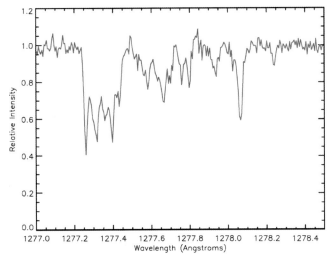

The High Speed Photometer

The High Speed Photometer (HSP) was built at the University of Wisconsin, under the direction of Professor Robert Bless and his HSP Investigation Definition Team. According to Bless, the photometer was designed to measure high-speed fluctuations of light from some of the most energetic objects in the universe. It had to be able to follow light fluctuations up to 100 000 samples per second.

During the first three years of the mission, the HSP functioned flawlessly, but, because of the spherical aberration and jitter, its science program was reduced. The spherical aberration in particular was quite damaging to HSP's observational capability. Because the mirror was spreading the light out in a halo, stellar images simply overflowed the aperture. Precise work with such a large image was impossible. However, the HSP was able to measure changes in light every 1/50 000 of a second.

The Fine Guidance Sensors

The Fine Guidance Sensors (FGSs) were built by Perkin-Elmer, now Hughes-Danbury Optical Systems. These sensors allow the telescope to acquire light from the guide stars it needs for accurate pointing. One of the FGSs also functions as an additional scientific instrument on HST, providing star positions that are ten times more precise than ground-based measurements. The Earth's atmosphere blurs star images taken from the ground, and consequently astronomers cannot measure their positions too accurately. High above the atmosphere, however, the seeing is fine, and from there astrometry can be done exceedingly well.

HST's Fine Guidance Sensors are ideal space-based astrometry instruments, and a scientific program is being carried out under the direction of a University of Texas astronomy team led by William Jeffreys. They are being used along with the European Space Agency's Hipparcos satellite data to measure positions and parallaxes of stars. Because spacecraft time is so limited, and because it takes months to achieve parallaxes with HST, Jeffreys and his team have to concentrate on objects of particular interest, either for their astrophysical characteristics, or for their function as fundamental distance calibrators. What the team loses in numbers of parallaxes, it more than makes up for in terms of accuracy of parallaxes: 'We have about a half a dozen stars in the Hyades that we're going to try and wring the best parallax out that we can get,' Jeffreys explained.

Hipparcos, on the other hand, did a faster survey, looking at more than 120 000 stars out to 11th magnitude. One of the goals for astrometry with HST is to achieve reliable parallax

Figure 1.21. Sample of data from the High Speed Photometer. It was taken while observing a star being occulted by the rings of Saturn. The star moved from the middle into the outer Crepe ring, and several obvious features are shown. (This data sample is discussed further in Chapter 2.) *(Robert Bless, University of Wisconsin and the HST Team)*

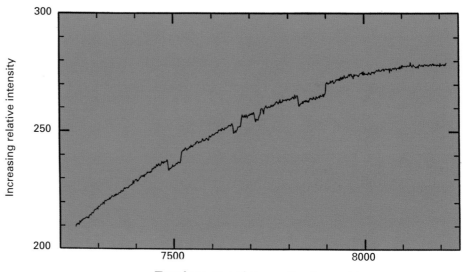

measurements for very faint stars and such difficult-to-measure stars as binaries and Cepheid variables. Other projects underway with the Fine Guidance Sensors are extrasolar planet searches around Barnard's Star and Proxima Centauri, as well as a study of the stars in the Hyades star cluster.

The FGSs work by knowing where to expect a guide star to be and then acquiring the star in their apertures. The astronomer designing an HST observation plan has to know the positions of guide stars in advance. Early on in the mission, no database existed that contained more than about half a million stars. So, a catalog was created containing the stars that could be used as guide stars by HST. One of the Astrometry Team's early tasks was to advise NASA on how to create that database. They built a prototype system in Texas, and then delivered the design for the guide star system now in use by HST's controllers.

COSTAR

The spherical aberration of HST's main mirror was 'fixed' by an ingenious optical instrument called the Corrective Optics Space Telescope Axial Replacement, or COSTAR. This instrument originally deployed five pairs of coin-sized mirrors in front of the Faint Object Camera, the Faint Object Spectrograph, and the Goddard High Resolution Spectrograph. The mirrors refocused the incoming light for these axial instruments, correcting for the spherical aberration. In 1997, the 'swap out' of GHRS and FOS for STIS and NICMOS left the Faint Object Camera as the only instrument using COSTAR.

COSTAR was built for NASA by Ball Aerospace Systems Division, under the direction of Ball Program Manager John Hettlinger and Space Telescope Science Institute's Holland Ford. More than just improved vision was at stake when Ball took on the COSTAR project. Ford and others described the engineering stakes as being very high. Not only was NASA sporting a black eye from the discovery of spherical aberration, but anyone who tried and failed to repair HST would share in the disgrace. Fortunately, Ball put an exceptional team to work on the problem, including one of their best opticians available, the late Murk Bottema. Due to the hard work put in by this team, COSTAR did not fail. At a press conference after the successful deployment of COSTAR, Ford explained the achievement: 'By putting all the light into a very small number of pixels, we can do in orbit the equivalent [with the FOC] of spotting a firefly in Tokyo from Washington, D.C. The resolution enabled by COSTAR is so good that if you used it to look at two fireflies nine feet apart in Tokyo, you could tell they were two fireflies from Washington.'

Observing with HST

The process of using HST for an observation is more complex than any other observing process in the history of astronomy. Hundreds of users have been assigned and have used HST time, but to do this they worked their way through what many have termed the 'most incredibly complex' process of proposal and observation allocation ever seen.

With ground-based facilities, if observers want observation time they apply to the observatory and describe the observations they want to make. A group called a Time Allocation Committee (TAC) reviews the request and either approves or rejects it. If it is accepted, the scientist can travel to the observatory, set up the equipment, with the help of a night assistant, and do the observations. Then the observer takes home photographs, computer disks, videotapes – whatever storage medium is needed. Final data analysis takes place at the observer's 'home' institution. Often, the observer stays home and monitors the observations via computer linkup, and has the data delivered.

In principle, the system is similar for HST, but there are some major differences. Observations with an observatory in low-Earth orbit can be an involved process. The orbital period produces an observatory 'day' of a little over 90 minutes. Because of this, the observations are often chopped into small segments. As a consequence of this difficult observing environment, the ground system must command many spacecraft functions, but these are not under the direct control of the observer.

An observer still applies for time on HST, but it is done in several steps. First, the observer applies for the time, describing the object to be observed, the instruments needed, and the scientific justification for the observation. The application is sent to the Space Telescope Science Institute via electronic mail. An HST Time Allocation Committee meets yearly to consider these proposals for using HST, and sends its recommendations to the Institute Director for final consideration. Eventually, time allocations are made, and observers are notified of their 'place in line' to use the telescope.

There are several categories of observers on HST: one group consists of the so-called 'guaranteed time observers', who were promised a certain amount of observation time in return for their work on developing the various instruments on board the telescope. There are also general observers, who apply for time in yearly competitions. One special group of observers no longer exists – the amateur group. However, when the HST Amateur Program was going strong, it was a popular one. Amateurs applied for a special block of time set aside for the use of the Space Telescope Science Institute Director (called Director's Discretionary Time). Their applications were considered by a special TAC set up by Jet Propulsion Laboratory's Steve Edberg and Space Telescope Science Institute's Peter Stockman to judge amateur proposals. Although some science did come from the amateurs, the program was dismantled in 1996 because the scientific gains did not balance the amount of effort needed to guide the amateurs through the entire proposal, observation, and reporting process.

Once observing proposals are accepted, they are encoded into a special set of instructions for the ground-based control systems to follow. In a typical command load, HST will be looking at something – Mars, for example. The next observation, which may or may not be planned to take advantage of the current slew position, comes up and tells HST to slew to the Orion Nebula. The program will contain guide star information for HST's star-trackers to find, and a command to start slewing to a position near where the guide stars are. If all pointing parameters are satisfied, i.e. if the object is not closer than 50 degrees to the Sun, or too close to the Earth's limb, or the Moon, and if the telescope is not close to (or going through) a region of strong particle radiation called the South Atlantic Anomaly, then HST can proceed to lock on to its guide stars and start the search for the object in question.

When HST is in position and the instruments are ready to go, it is time to do the observation. Some observations take only a few minutes; others take several hours. Since HST is in orbit, hours-long observations continue over many orbits. Each time the telescope moves out of sight of the object, it loses its lock on that object. As soon as HST comes back around the Earth, it must relocate its target and continue the observation. If the observer has chosen an object near the poles of the Earth, an area called the 'continuous viewing zone', then the telescope can observe for longer periods of time.

At the end of the observation, data are saved by a tape recorder on board the telescope. At some point the data are sent to Earth, where they are received at a ground tracking station at White Sands, New Mexico. The data are then relayed through Goddard to the Space Telescope Science Institute, where they are usually released to the observer within a few hours or days.

Unlike a ground-based observatory, very few scientists sit around Space Telescope Science Institute watching their data come down from the telescope. Usually the astronomers are aware of their scheduled observation times, but wait out the observations at their home institutions. However, some observers do come to the Institute, and they often work with people whose job it is to monitor the observations and take action if severe problems arise.

Observation with HST is a lengthy process. For most observers, more than a year will pass between the time of first proposal submission and the time the observation is actually

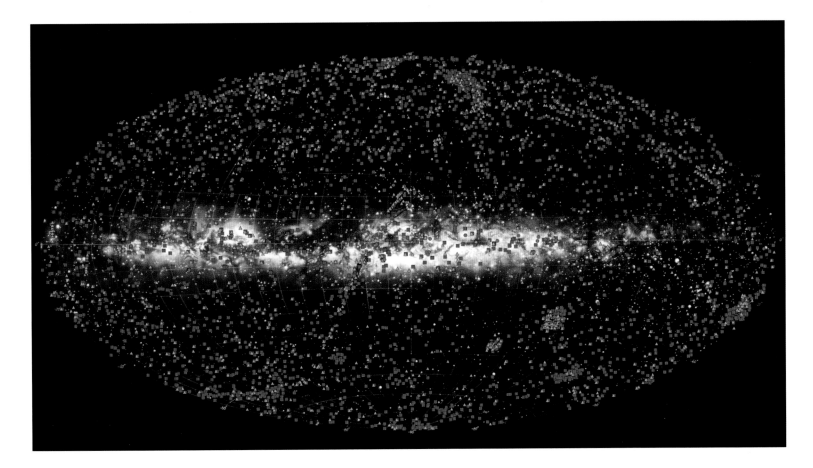

Figure 1.22. HST's view has encompassed nearly the entire sky. Each square on the map represents a viewing 'hit point' for the Hubble Space Telescope. *(Zoltan Levay, STScI)*

executed. Contrast this with the time frame of several months needed to complete an observation at a ground-based observatory, and one can see why observing with HST is not a simple undertaking.

Despite the complexity of the observation process – and the inherent frustrations of applying for time and using the telescope – increasing numbers of astronomers vie for time on HST. They follow it through, and, if all goes well, they are rewarded with information that at least one scientist frankly characterized as 'wonderful data, but was sometimes painful to get'.

Within the community, experienced observers cannot help but compare the differences in operation between HST's hierarchy and that found at any ground-based institution. According to Ray Weymann of the Observatories of the Carnegie Institution of Washington, the procedure may or may not be worse than a ground-based facility. 'It's a complex procedure,' he says. 'People who are very familiar with the details say that if you know the ins and outs, if you're very close to Space Telescope, there are ways of doing things which can save you a lot of trouble.'

It is important to realize that many of HST's early problems should be categorized as 'growing pains'. In response to user concerns, the Institute organized a complete overhaul of the way it does business with the telescope's users. Some of this overhaul was long overdue, but some of it is due to a change in philosophy prompted by Robert Williams when he took over the Director's job in 1993.

Quite aside from the HST user community, there remains a high public interest in the telescope's operations. The Space Telescope Science Institute maintains an active Office of Public Outreach, which supplies press announcements of HST discoveries to members of the media. The educational community is served by the same office, which makes materials available to teachers on request.

There is no doubt that HST has taken its place among the premier observatories of the world. From observations of faraway quasars, to comet crashes in our own cosmic backyard, scientists have come to recognize the value of the telescope. To be sure, it is still very much a political animal. In a day and age when governments are looking for any way possible to cut spending, HST might even be termed a 'Big Science' luxury. The question that everyone can ask is whether using the telescope is a worthwhile endeavor. Even a casual look at the science it has done during its years on orbit tells us that it is.

2

HST and the Solar System

On a dark hillside overlooking Padua, Italy, in the year 1610, astronomer Galileo Galilei studied the sky through a crude telescope. He was gazing eagerly at a bright object that had tantalized his curiosity for years. Much to his surprise, he found that this object – the planet Jupiter – had four companions that seemed to circle it much as he had observed Mercury and Venus to circle the Sun. Over a series of nights, he watched Jupiter's tiny companions move in an intricate dance near the planet, and he concluded that they were a family of moons in Jovian orbit. To him they seemed to be an analog of the Solar System. In time, Galileo focused his gaze on Saturn, and saw something around that planet as well. Perhaps, he reasoned, they were moons as well. Through his simple instrument they looked more like ears. While he often speculated what these things were, he never did resolve the mystery of these Saturnian companions.

Galileo was not the first to notice movement in the heavens. During the centuries before his observations, others had studied the heavens, and noted the actions of so-called 'wanderers'. These point-like objects seemed to move at will across the sky. People went to great efforts to chart and predict them. While earlier astronomers, such as Aristotle and Ptolemy, knew of at least five planets by their motions, it did not occur to anyone that these wanderers – these planets – were worlds in their own right. Galileo's journal entry tagging these planets as other worlds permanently changed humanity's views of the Solar System.

In her starry shade
Of dim and solitary loveliness,
I learn the language of another world.

Lord Byron

I would like to show you how to fall in love with a planet.

Robert M. Powers

Pre-HST exploration

Four decades of active space exploration continue to change our view of the Solar System. Like explorers of earlier centuries, robotic spacecraft have scouted out the territory, reporting back a tantalizing array of first impressions. Through the electronic eyes and ears of our spacecraft, we have seen an incredible variety of planetary surfaces, atmospheric structures, and moons scattered throughout the Solar System.

In the 1960s the first robotic explorers went to the Moon – spacecraft with names like Ranger and Rover. Mars was the next target, studied by a series of Soviet Mars probes and the US Mariner craft. They prepared the way for the Viking landers in the mid-1970s, and a series of late-20th-century explorations that began with the landing of Mars Pathfinder in 1997 and orbital mapping carried out by the Mars Global Surveyor beginning in 1998. One Mariner craft studied Mercury; two Pioneer Venus spacecraft, a series of Venera spacecraft, and Magellan went to Venus; Pioneer 11 and 12, and Voyager 1 and 2, went on fly-by missions that included all the outer planets except Pluto; the Galileo spacecraft looked at Venus and Earth on its way to an extensive mission at Jupiter. Finally, Ulysses – the solar–polar mission – is examining the solar wind from various vantage points around the Sun.

Astronomers have even received their first close-up studies of asteroids and comets, probing Gaspra, Ida (and discovering Ida's small moon Dactyl), and Mathilde, as well as comets Halley, Giacobini-Zinner and Grigg-Skjellerup. Earth-approaching asteroids and the Shoemaker-Levy 9 impacts on Jupiter spurred scientists to mount a systematic search for small bodies in the inner Solar System. Others are concentrating on the population of small bodies out among the gas giant planets and beyond. It has been a magnificent four decades of exploration along the shores of the cosmic ocean, and, as the late astronomer Carl Sagan wrote in *Cosmos*, 'we have waded a little out to sea, enough to dampen our toes, or at most, wet our ankles.'

So, you might ask, what is left to study in the Solar System? In a word: everything! What we have gained over these past decades of planetary exploration is a mere nodding acquaintance with the Solar System and some interesting theories on how it may have formed. What we lack is the depth of understanding that prolonged study can give us. While we have been able to categorize the forces that shape solid bodies in the Solar System – cratering, tectonism, and volcanism of all types – we are still learning about the dynamics of large-planet atmospheres, the plasma physics of comet tails, and the true makeup of the many families of asteroids that inhabit the spaces between the planets. The Solar System we study today contains many tantalizing clues to its beginning, and those clues can lead us to a greater understanding of the forces of starbirth at work throughout the cosmos.

The earliest cosmologies held that the Earth was at the center of the Solar System, if not the universe itself. That idea was replaced by the concept of a Sun-centered universe, which was in turn replaced by the current theory of an expanding universe with no center. Our thoughts on the origin of the Solar System itself have gone through similar changes. Not many years after Galileo Galilei's observations, philosopher René Descartes tackled the genesis of the Sun and the planets. His was a systematic and scientific approach based on precise observations rather than biblical suppositions or unscientific 'theories'. Descartes theorized that vortices formed in a primordial gas cloud, and that they were responsible for the formation of the planets. His early theory did not stand up for very long, and the next person to try to explain the origin and evolution of the Solar System was a French nobleman – the Comte de Buffon. He suggested that the planets formed when a comet plowed into the Sun, the impact causing it to eject material that later became the planets. Buffon was the first of many proponents of what we now call 'catastrophe' theories of Solar System evolution. Some of these theories postulated the collisions of stars as the seed activity for planetary formation, and that the action of one star merely passing by another would be enough to start the planetary birth process.

In the 18th century, philosopher Immanuel Kant advanced a theory that the Sun and the planets formed in one continuous process. The idea was fine as far as it went, but it needed a bit more definition. Mathematician Pierre Simon de la Place took it one step farther. *He* saw the birth place of the Solar System as a spinning gas nebula, and this theory is the prototype of the creation story astronomers tell today. The Solar System's birth place was most likely a cloud of hydrogen and helium gas, plus a good supply of heavier elements ejected from the explosion of massive stars (supernovae). The cloud began to contract, possibly triggered by shock waves from a nearby supernova. The waves of contraction compressed the material in the primordial nebula, and the compression caused the cloud to heat. Fluctuations may have set the nebula spinning, slowly at first, but speeding up later as gravitational attraction began to pull clumps of material together. As the spinning continued, the nebula flattened out and took on the orbital characteristics of the proto-Solar System.

The bigger agglomerations attracted more and more material to themselves. The largest clump became the proto-Sun. Dust grains trapped in the nebula began to stick together to form planetesimals, which collided and merged with one another to become the proto-planets.

At some point, the compressional heat in the proto-Sun was high enough to trigger nuclear reactions. When that happened, the proto-Sun became a star, flaring fiercely in its newborn youth, sending bipolar jets out across the light years as a sort of cosmic beacon heralding starbirth. Winds from the newly born Sun blew the remaining gas cloud away from the inner Solar System, leaving only the Sun and four rocky bodies that were to become airless Mercury and atmosphere-rich Venus, Earth, and Mars. The larger planets, with their higher masses and gravities, managed to hold on to hydrogen and helium atmospheres, becoming Jupiter, Saturn, Uranus, and Neptune. The leftovers of this birth process populate the Solar System as comets and asteroids, and the many moons that orbit the planets.

This modern theory of Solar System formation is far from perfect. It leaves many unanswered questions about the origin of the primordial cloud and the rate of formation of objects within it. The theory does not completely explain the existence of the asteroid belt between Mars and Jupiter, or the size and extent of the Kuiper Belt and Oort Cloud of cometary nuclei in the outer Solar System. Still, it is the best theory we have. The only way to prove it right or wrong is to continue our Solar System exploration efforts, and to look for planetary systems around other stars.

Researchers have classified the worlds of the Solar System in a number of ways: terrestrial worlds and gas giants; planets with atmospheres and those without; worlds with moons and those without; asteroids, near-Earth asteroids, comets, Sun-orbiting comets, Jupiter-orbiting comets, proto-comets in the outer belts; and a host of others.

The Earth is the benchmark against which we compare all other planets because it is where we originated. Earth's nitrogen/oxygen atmosphere, for example, seems to be unique in the Solar System. Mercury has no air, and Venus has a poisonous carbon dioxide and sulfur dioxide atmosphere. Mars has a thin carbon dioxide atmosphere, and it may have had a heavier blanket of gases in the distant past. The gas giant planets have atmospheres abundant in hydrogen and helium, with small amounts of methane and ammonia. The thickness of Pluto's diaphanous nitrogen and carbon monoxide atmosphere varies with its distance from the Sun. We know that our own atmosphere was mainly carbon dioxide in the early days of the planet, and that it has been modified by continued volcanic outgassing and other processes such as life since then. However, Venus and Mars also show evidence of volcanism, so what is the difference? Certainly the presence of life and water on Earth make some difference in the evolution of Earth's atmosphere, but how much? And how do we know for sure? What processes are at work on other planets? Could those processes lead to life on those worlds as well?

To answer these and many other questions, we aim our research instruments at the other planets. One useful way to do this is by means of orbiting spacecraft such as Magellan for Venus, and fly-by spacecraft such as Pioneer and Voyager. Unlike long-term missions such as the Mars Pathfinder lander or the Galileo explorations of Jupiter, many of the probes sent on planetary voyages have been fly-bys on fixed trajectories – capturing brief moments in the life of a planet before moving on. What we receive are planetary 'snapshots' – images of worlds frozen in an instant of time. Unfortunately, there are only a few new planetary probes being built to be sent out on voyages of discovery, and for long-term studies, snapshots are not enough. The worlds of the Solar System are dynamic places. They bear watching constantly because they change constantly. To truly understand planetary change, we need to make more observations of our Solar System neighbors over long periods of time.

As we await the launch of more planetary exploration vehicles, HST provides us with a way to study many Solar System worlds. The main advantage that HST has over ground-based and fly-by type missions to the planets is that it can look at an object over a long period of time through broadband filters and can carry out high-resolution spectroscopy of the other members of the Solar System.

Observing planets with HST is a tricky business. Ironically, these objects – which are Earth's closest 'cosmic neighbors' – present the telescope with its greatest observing challenges. Like any Earth-based telescope, HST has to track its moving targets. Compared to stars and galaxies, however, planets move relatively rapidly against the backdrop of the sky. Planetary scientists using HST must always account for these motions when they plan their observations, and, of course, the telescope's own orbital motions have to be added into the observing calculations. The current method of acquiring and tracking Solar System bodies with HST involves a combination of 'ambush mode', guide star hand-offs (where the telescope uses a succession of guide stars to help keep it pointed accurately as it orbits the Earth) and a slow scan rate that matches the movement of the object.

Despite the difficulties involved in observing planets, HST has compiled a respectable record of Solar System exploration: since 1990, it has studied the atmosphere and surface of Mars, imaged the cloud tops of Venus, studied the clouds and aurorae of Jupiter, tracked a

Figure 2.1. A panoramic view of Ares Valles, taken by the Mars Pathfinder lander, July 1997. (NASA; JPL; Caltech)

storm through Saturn's cloud decks, given us a 'ringside' seat as Earth crossed the plane of Saturn's rings, discovered new moons and mapped old ones, imaged and mapped a selection of asteroids, imaged and taken spectra of a dynamically exciting collection of comets, studied the atmospheres of Uranus and Neptune, identified proto-comets in the Kuiper Belt, and mapped the surface of Pluto. Not a bad record for an Earth-orbiting explorer!

Hubble's dynamic Solar System

Mars

People have always been fascinated by Mars. It stands out in the night sky as a reddish-tinged, unwavering point of light. If we look at it through an amateur telescope (with, say, an 8 inch instrument or larger aperture), we can just make out faint markings on the Martian surface. Over the course of a Martian year, we can watch the polar caps grow and shrink with the change of seasons. We know Mars has an active atmosphere because planet-wide storms sometimes blanket the surface with dust. For clearer views of the planet, we need larger telescopes. However, even the largest ground-based instruments cannot resolve small details on the Martian surface.

The most meticulous explorations of Mars were conducted in the 1970s, when a series of spacecraft from the USA and the then Soviet Union imaged the planet. Two Viking landers settled onto the surface of the Red Planet in 1976, and returned a steady stream of data about atmospheric and surface conditions. Recent attempts at exploration with two Phobos craft and a Mars Observer mission ended in disaster as contact with all three probes was lost. More Mars missions are underway to continue the studies begun more than 30 years ago.

Why all the interest in Mars? Certainly the planet resonates in our collective unconscious as a place to explore. Some Mars enthusiasts point to the planet as a future home for Earth colonists. Others see in Mars a chance to study a planet similar to the Earth in many ways. A great deal of scientific interest in the Red Planet focuses on its atmospheric conditions and surface characteristics. Understanding what happens on Mars and comparing it to the conditions we experience on Earth is a strong motivation for Mars research. For those who want the Red Planet to be a second home to humanity, a knowledge of climate and terrain would be essential. To form the most accurate picture of changes on Mars, however, we need to monitor the planet over many years.

There are many questions we seek to answer about Mars, many of them aimed at finding out if the planet was warm and wet during its early days. If it *did* have more water in its infancy, the next question to ask is whether it formed with the same amount of water as the Earth did, or with less. If we assume that it started out with the same amount as the Earth, then we must ask what happened to cause Mars to lose most of that water. The alternative is to assume that it started out with much less and then to try to find out why.

Either way, we are faced with the certain knowledge that some amount of water once flowed on Mars because we see evidence of it all over the surface. Lengthy flow channels with tributaries spread across the dusty plains, looking remarkably similar to dry riverbeds on

Earth. Ancient, rock-strewn plains reveal a landscape tortured by the catastrophic rush of water across the surface. Where was the water flowing to? And where is it now? It is possible that some small fraction of the original water supply is locked underground in permafrost, but it seems certain that a great deal of it escaped into space. The question is, how do we go about proving that this happened?

One possible way is to look in the upper atmosphere for the presence of *deuterium*, an isotope of hydrogen. To form deuterium, you take the nucleus of a hydrogen atom (which contains one proton) and add a neutron to it. Since hydrogen is the main element in water (water is two atoms of hydrogen and one of oxygen: H_2O, you might expect to find some deuterium in the water in the upper atmosphere.

Imagine if the early Earth were to lose massive amounts of water to space in a runaway evaporation cycle. The water, in the form of vapor or gas, would rise high up into the atmosphere. There it would break apart – dissociate – into hydrogen and oxygen. Hydrogen, being very light, would escape into space rapidly, whereas any deuterium, which is heavier, would escape slowly, and much would be left behind. Now assume that millions of years go by as this process takes place. A group of aliens comes by, exploring the Solar System. In their travels, they find this dry and desiccated planet, pockmarked with craters. It looks as if it used to have water, because its surface is criss-crossed with fossilized channels, dry riverbeds, and flood plains. The big question is: why is it that way?

To find out, they start to measure the atmospheric water content. They find a little water vapor and lots of deuterium. That deuterium content tells them that a large amount of hydrogen, from water, escaped into space over time, leaving behind the deuterium. The more deuterium found in the upper atmosphere, the more hydrogen must have escaped. If our aliens calculate a deuterium-to-hydrogen ratio, and work backwards from that ratio, they might deduce that this desert planet once supported huge oceans and teemed with life. Could the same be done for Mars?

For many years, planetary scientists *have* searched the upper atmosphere of Mars in the same manner as our hypothetical aliens, looking for the spectral signature of deuterium. They have looked at the data gathered by all of the missions to Mars in an effort to gain some understanding of the planet. However, Mars exploration did not stop when the Viking 1 spacecraft stopped sending data in 1982; in the years since then, planetary scientists have studied the planet from ground-based observatories, from NASA's Kuiper Airborne Observatory, and by analyzing the makeup of a peculiar set of meteorites found on Earth that are thought to have come from Mars. These studies have allowed planetary scientists to make some very sophisticated estimates of the existence of water, deuterium, ozone, and other gases in the Martian atmosphere. For example, based on theoretical models of hydrogen escape, planetary scientists have come up with a ratio of deuterium to hydrogen on Mars which is five times that of the Earth's. Some estimates of water loss point to Mars losing a layer of water which was somewhere between 3 and 27 meters deep! Clearly there is a lot of work to be done as we resume our *in-situ* studies of this planet.

A team led by French astronomer J.L. Bertaux of the Service d'Aéronomie du CNRS, used the Goddard High Resolution Spectrograph to obtain a short exposure spectrum of the

Fig 2.2. A color composite of images of the Syrtis
Major region on Mars, taken with WF/PC-1. *(Philip
James, University of Toledo; Steve Lee, University
of Colorado; NASA)*

Figure 2.3. Mars, taken by WF/PC-2 on March 21, 1995. *(Philip James, University of Toledo; Steve Lee, University of Colorado; NASA)*

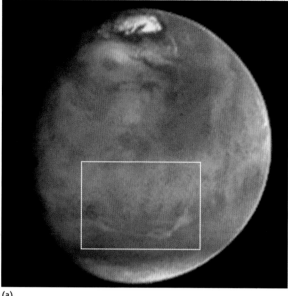

(a)

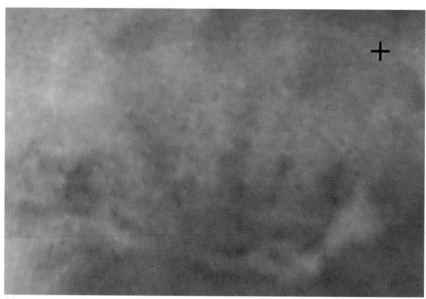

(b)

Figure 2.4. (a) On June 27, 1997, WF/PC-2 imaged a dust storm in the Valles Marineris, about 1000 km away from the Pathfinder landing site. **(b)** is a closeup view of the storm. The + marks the pathfinder landing site. *(Philip James, University of Toledo; Steve Lee, University of Colorado; NASA)*

Martian atmosphere. The team reported a possible detection of deuterium in the Martian upper atmosphere in the form of a spectral signature called the 'D Lyman-alpha line'. This finding showed that planetary scientists could use the telescope's spectrographs to study the Martian atmosphere. In recent years, other observations have concentrated on measuring the abundance (amount) of carbon dioxide in selected parts of the Martian atmosphere. Others have been searching the upper atmosphere for deuterium and comparing the abundances to amounts suspected to exist in the lower atmosphere. If these observations can be continued and expanded, the next step will be to apply the data to determining the timing of the events that triggered the catastrophic loss of water on Mars.

Other changes in the Martian atmosphere are the main interest of a group of researchers who have spent several years devoting their HST time to an ongoing study of Mars. Philip James of the University of Toledo, the late Leonard Martin of Lowell Observatory, the University of Colorado's Steven Lee, and Todd Clancy of the Space Science Institute in Colorado, instituted a series of observations to measure the changes of albedo (reflectivity) of various surface features on Mars, as well as to study the opacity of the Martian atmosphere and to take measurements of the seasonal and interannual variations in the ozone distribution. Apparently, the creation of water vapor is part of a chain of chemical reactions that remove ozone from the Martian atmosphere. HST's ability to image Mars in the visible and near ultraviolet allows the team members to measure the absorption of ozone, which, as it does on Earth, plays a role in the upper atmosphere. Of particular interest is the part ozone plays in atmospheric changes over the Martian polar regions. The team also wants to learn more about how the polar ice caps on Mars change with the seasons. The preliminary results of the Mars team's work suggest that the amount of ozone in the Martian atmosphere changes seasonally with latitude. There is also evidence for strong ozone absorption in the north polar region of Mars during late winter.

In a striking sequence of images released by James and his team since 1990, some features familiar to Mars-watchers show up in sharp detail. In Figure 2.2, for example, the large dark area shaped like a shark's fin is the Syrtis Major Planitia. The surface material here is probably a coarse, dark-colored sand. Prevailing winds blow the dust into the brighter-looking Arabia Planitia to the west. To the east of Syrtis lies Isidis Planitia, an impact basin formed more than two billion years ago when a meteorite slammed into Mars. In Figure 2.3, the Ascraeus Mons volcano stands out in sharp relief above the high, thin clouds at the planet's limb.

The planet-wide dust storms that so fascinated observers back in the 1970s do not seem to be recurring with any regularity. Steven Lee monitored the region around Syrtis Major, which is one of the most variable regional albedo features on the planet. Albedos change when winds

(a)

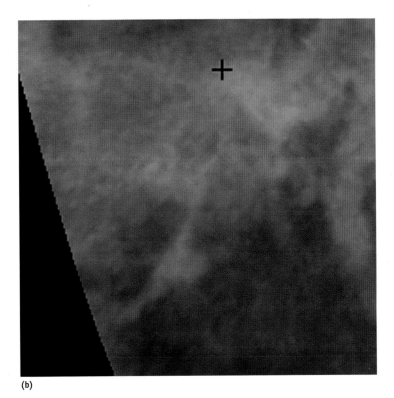

(b)

move dust and sand around on the surface, and this process of wind transport may be responsible for many of the variable albedo features seen on Mars. By early 1994, Lee and his colleagues had seen no major dust storms in their HST data, although a major dark albedo feature called Cerberus had disappeared under the shifting dust and sand of the planet. This was not evidence of a global storm, yet, if Viking data were to be believed, the team should have seen major changes on the Martian surface if a dust storm *had* occurred. In at least 12 separate observations of the region, little variability was seen. Ironically enough, it appears that a storm may have blanketed large regions of the planet during early 1994, but, as with many another project, the team did not have any HST observations scheduled at that time. The lack of globe-girdling dust storms in the style of those that blanketed the planet during the Mariner and Viking explorations has been puzzling.

More recent Mars work *has* resulted in the remarkable series of photographs seen here showing small-scale storms and seasonal changes on the surface of Mars as well as in its atmosphere. Clouds appear quite regularly above the Martian surface, due in large part to an atmosphere remarkably free of major dust activity. If there was a lot of dust in the Martian air, it would be absorbing sunlight and heating the atmosphere. Mars' atmosphere is quite cool, which causes water vapor to freeze out and form the thin, ice crystal clouds seen in the HST images. This long-term observing program has changed our view of the Martian climate, showing us a Mars atmosphere that is cooler, drier, and clearer than it was during the 1970s when the Viking spacecraft gave us our first long-term look at the planet.

In 1997, HST Mars observations of small-scale storms took on added meaning as early warning reports for the weather facing the Mars Pathfinder Mission. The week before the spacecraft's July 4, 1997, landing, HST images showed a dust storm kicking up in the Valles Marineris, some 1000 kilometers south of the landing site. High, thin cirrus clouds hovered over the landing site, with thicker clouds further north. The week after the landing, the southern storm had dissipated, but a new storm had appeared about 2000 kilometers north of the landing site, near the Martian north polar cap. It appears that the Pathfinder slipped down to the surface between storms, during a time when the regional distribution of clouds and dust was changing. Long-term studies of the planet's atmosphere continue with HST in an effort to track widescale seasonal changes in the planet's weather, and in support of the ongoing Pathfinder and Global surveyor missions to Mars.

Figure 2.5. (a) A small-scale storm raging near the northern polar cap on Mars on July 9, 1997, a week after the Pathfinder landing on Mars. **(b)** A closeup of the Valles Marineris region, with the Pathfinder landing site marked +. *(Philip James, University of Toledo; Steve Lee, University of Colorado; NASA)*

Venus

No less striking in our skies than Mars is the lovely, cloud-covered Venus. For years, observers gazed at it, wondering what lay below that yellow blanket of clouds. Radar observations from Earth, as well as the immensely successful Magellan mapping mission, revealed the planet to be a hellish place of volcanic activity, sulfuric acid rains, and tectonic upheaval.

Early in 1995, University of Colorado planetary scientist Larry Esposito used HST as a sort of interplanetary weather satellite to study Venus. His observations focused on the thick blanket of clouds that covers the heavily volcanic surface of Earth's so-called 'sister planet'. You might think that Venus is an unlikely target for HST observations, and for much of its orbit it is. This is because the planet always appears very close to the Sun, and HST is not allowed to

Figure 2.6. This WF/PC-2 image of Venus was taken on January 24, 1995, when Venus was 70.6 million kilometers from Earth and at its greatest elongation from the Sun as seen from Earth. False color has been applied to the image to enhance the cloud features. *(L. Esposito, Laboratory for Atmospheric and Space Physics, University of Colorado, Boulder; NASA)*

let its gaze wander too close to the Sun. For a few weeks each orbit, however, Venus is just far enough away from the Sun (as seen from Earth) that HST researchers can risk sneaking a peek at it. On January 24, 1995, HST took imaging and spectroscopic observations aimed at studying the cloud patterns on Venus. Figure 2.6 is HST's view of Venus.

At ultraviolet wavelengths, specific patterns in the Venus cloud tops stand out – particularly a large, horizontal, Y-shaped feature near the equator. Other spacecraft, sent to study Venus in the past, reported this peculiar formation, which may be a sort of 'wave' in the atmosphere, analogous to high and low pressure cells in Earth's atmosphere. The polar regions of the cloud deck appear bright, and could be covered with a haze of small particles. Dark regions in the clouds show the location of increased sulfur dioxide levels near the tops of the clouds.

The heavy cloud blanket on Venus is very different from Earth clouds, which are largely water vapor. On Venus, the clouds consist mostly of sulfuric acid, and it seems that the atmosphere was deluged with a sort of 'sulfuric acid rain' triggered by a volcanic eruption in the late 1970s. It appears that the atmosphere is still recovering from that eruption (there is much less sulfur dioxide in the dark regions now than there was in the early 1980s, for example), and researchers hope to use the telescope to study the Venus atmosphere the next time the planet is far enough away from the Sun to permit HST a safe view.

The Jupiter system

Nearly 400 years after Galileo turned his telescope to the Jovian system, the planet Jupiter became the subject of worldwide attention. Members of the public as well as the scientific community were about to witness a rare and spectacular sight – the collision of a comet with a planet. For one group of astronomers crowded into a control room at the Space Telescope Science Institute on a warm July evening in 1994, it was the end of a feverish year of preparations. They were waiting and watching as 21 fragments of Comet Shoemaker-Levy 9 crashed into the swirling clouds of Jupiter. It was an historic first for planetary scientists – the chance to see just what sort of changes would take place when pieces of a comet interacted with a gas giant planet.

The astronomers – part of an extended network of scientists around the world – would not actually see the pieces of Shoemaker-Levy 9 plunge into the Jovian cloud tops. Instead, everyone gathered at the Institute had pinned their hopes on the HST to capture both visible and ultraviolet views of the aftermath of the collisions. At the same time, ground-based astronomers used infrared instruments to measure the energy given off as each comet fragment was consumed in a huge fireball.

Shoemaker-Levy 9 – the famous 'string of pearls' comet discovered by the team of the late Gene Shoemaker, his wife and co-researcher Carolyn Shoemaker, and their colleague David Levy – mesmerized the world from the moment of its discovery. For years, the trio of observers had been doing standard searches for comets and asteroids from Palomar Mountain in California. On the night of their discovery, they were ready to give up observing because it was starting to cloud up. They decided to use up some defective film before stopping work for the night. Later on, Carolyn Shoemaker was looking at the film through an instrument called a stereo microscope.

In David Levy's words, it was an electrifying experience. 'Suddenly she [Carolyn] sat up straight in her chair and looked more intently [at the image on the instrument],' he said. 'Then she looked up at us and she said, "I think I've found a squashed comet!"'

The other two took turns looking through the stereo microscope and each saw what had so surprised Carolyn – a bar of cometary light, with tails leading off to the upper right. There was something 1 arcminute long that looked like a cometary coma. Spreading out toward the upper right were little tails, maybe five or six of them.

The trio reported their find and it was announced the following day as Comet Shoemaker-Levy 9, later renamed Periodic Comet Shoemaker-Levy 9. There were approximately 20 objects, moving along in formation, headed for a 6-day-long July 1994 collision with Jupiter.

The largest 'pearl' in the first HST image measures only about 5 kilometers or less across. The objects originally came from a larger comet that was torn apart by a very close approach to Jupiter in the summer of 1992.

A special HST campaign was put together, with dozens of investigators from around the world participating in the observations. HST watched the fragments until 10 hours before the impacts to see if the largest pieces would break up. These observations indicated that each of the largest 'pearls' in Shoemaker-Levy 9 went into Jupiter whole, trailing a stream of dust as they went.

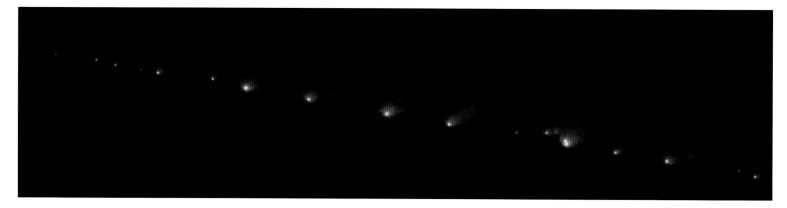

Figure 2.7. WF/PC-2 view of Periodic Comet Shoemaker-Levy 9 in January, 1994. The mosaic image consists of two WF images and one PC image; 20 'comet pieces' are shown. *(NASA; ESA)*

Figure 2.8. WF/PC-2 (in PC mode) image of Jupiter, showing multiple impact sites in July, 1994. Several sites are visible in a row, including the star-shaped 'H' site (below the Great Red Spot) and extending to the 'D/G' site near the limb (bottom). *(Hubble Space Telescope Comet Team; NASA)*

The cometary impacts affected the Jovian magnetosphere. When the comet first entered the magnetosphere four days before impact, HST saw strong emissions of ionized magnesium at the comet. During the actual impact events, the so-called K fragment stirred up unusual auroral activity over Jupiter's poles. This occurred about 45 minutes after the K fragment entered the atmosphere. The auroral features were well outside the regions where aurorae are usually seen on Jupiter (as we shall see below). The K fragment impact produced energetic, ionized particles that traveled along magnetic field lines into the upper atmosphere of the planet. When they came back down, these particles caused atmospheric gases to glow brightly in ultraviolet light.

For planetary scientists, the chance to watch a series of impacts in almost 'real-time' was a once-in-a-lifetime experience. Using HST to do it added a high-resolution component to an already exciting scientific event.

Indeed, the Great Comet Crash of July 1994 spurred planetary scientists to re-think their assumptions about impact events and collisions in the Solar System. Since the impacts, a group of astronomers has been searching for Earth-crossing asteroids and small comets that could pose a similar threat to our planet. Many Voyager images of icy moons that exhibit crater chains are being re-examined as new evidence that other comets like Shoemaker-Levy may have existed in the past.

Aside from the exciting momentary experience of using a comet as a probe of a giant planet's upper atmosphere, Jupiter continues to present a treasure trove for long-term study by HST. A variety of HST research programs focus on the turbulent atmosphere of the planet, the interaction of that atmosphere with Jupiter's magnetic field, and studies of the Jovian satellites.

From the spectacular Voyager 1 and 2 images, most of us are familiar with Jupiter's moons, its cloud belts and zones, and the Great Red Spot – which has been part of the Jovian 'cloud-scape' for at least 300 years.

Figure 2.9. A WF/PC-2 image of Jupiter, showing aurorae produced 45 minutes after the K impact. The site at the bottom left produced energetic particles which traveled along the Jovian magnetic field (marked by arcs at left) and slammed into Jupiter to produce the aurorae (bright regions at upper end of arcs). Jovian aurorae do not normally occur at these locations. The normal aurorae are visible at the north and south poles (see Figure 2.11). *(John T. Clarke, University of Michigan; NASA)*

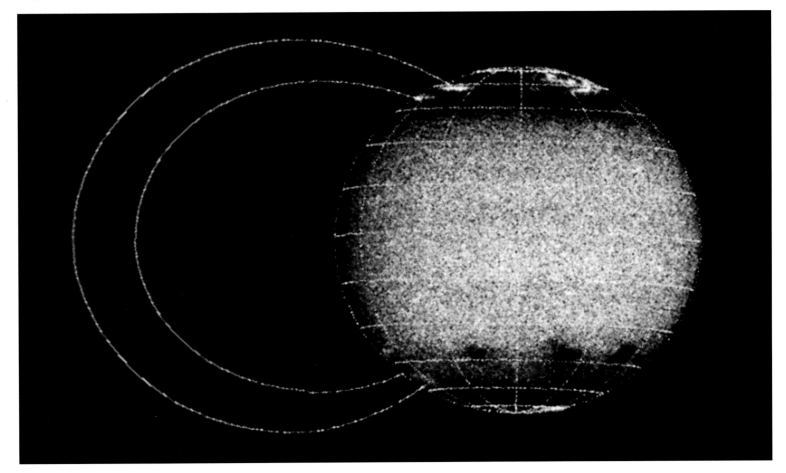

Figure 2.10. This February 13, 1995, HST image reveals surprising detail in the variety of storms and wind patterns in the Jovian cloud belts and zones. *(R. Beebe, New Mexico State University; NASA)*

However, studying Jupiter with HST reveals a world very different from the gas giant visited by the Voyagers in 1979. For one thing, the Great Red Spot continually changes its color. Also, the cloud belts constantly evolve and mutate, and multitudes of smaller storms have raced across the face of the planet and disappeared.

As useful as the Voyager work is for showing the scale and extent of features in the Jovian atmosphere, it is hard to get a feel for Jupiter's true nature from a series of 'snapshots'. HST's contribution to the study of Jupiter is an ability to deliver high-resolution images of the planet every 1.5 hours, to take ultraviolet spectra of the planet, and to give a more consistent feel for what the planet's upper atmosphere does over the long term.

Jupiter's atmosphere is its most striking feature. Belts and zones seem to divide the planet's appearance into colorful cloud strata. Storms of all shapes and sizes reel across the top of the atmosphere as the clouds blow past each other in opposite directions. This constant seething motion produces incredible wind shears and displacements in the cloud zones of the planet. These displacements were of great interest to the Galileo science team members as they prepared a special probe to drop into the Jovian atmosphere. Reta Beebe, from the New Mexico State University in Las Cruces, New Mexico, obtained images of Jupiter using HST's Wide Field and Planetary Camera, and did a series of observations of the planet before the Galileo spacecraft released a probe into the atmosphere in December, 1995.

In addition to long-term camera studies of the Jovian cloud tops, HST's spectrographs were used to scan Jupiter, gathering data on the aerosol hazes in the upper atmosphere of the planet and probing the limb darkening over certain latitudes. In addition, interactions between particles in the atmosphere and the planet's huge magnetic field are visible to the telescope in ultraviolet light. The Jovian magnetic field is the strongest and most extensive of any of the planets in the Solar System. Jupiter's magnetosphere is generated by the planet's fluid, electrically conducting interior, and it extends out about 2 million kilometers, well past the orbit of its moon Callisto.

Although HST cannot see Jupiter's magnetic field, it does study the effects of the field on particles in the upper atmosphere. Jupiter's magnetosphere traps electrons along lines of magnetic force, as well as sulfur ions from the tiny moon Io. Radiation from these electrons results in radio emissions, and aurorae occur over Jupiter's polar regions when the ions slam into the atmosphere. These aurorae were first seen in ultraviolet imaging from the Voyager 1 spacecraft in 1979. Since then, they have become one of the most studied Solar System phenomena by astronomers using HST. At stake here is a finer understanding of the small-scale details of Jupiter's electromagnetic environment.

University of Texas astronomer Larry Trafton used HST's Faint Object Camera and the Goddard High Resolution Spectrograph to look at Jupiter's northern polar region, while the Ulysses spacecraft simultaneously measured Jupiter's x-ray and trapped-particle emissions. Trafton looked at the ultraviolet light output of one small region within an aurora. To do that, the instruments were 'tuned' to observe hydrogen gas (H_2) emissions. When H_2 molecules are excited by interaction with a magnetic field, they radiate ultraviolet light. By focusing the spectrographs on the auroral regions and measuring the strength of the emissions, Trafton and his colleagues found that the gas was heated to a temperature of 1200 kelvin (about 927 degrees Celsius) in the aurora. Figure 2.11 is HST's ultraviolet view of the Jovian aurora.

HST's sharp view of the Jovian aurorae has allowed other astronomers to map the phenomenon against the immense magnetic field of the planet. Astronomer John T. Clarke of the University of Michigan used the telescope to record daily changes in the intensity and motions of the aurorae. Surprisingly enough, the northern and southern aurorae mirror each other on Jupiter, much as they do on Earth.

Changes in the brightness of the aurorae occur over the course of Jupiter's 10-hour day, probably in response to the compression of the magnetic field on the sunward-facing side of the planet. One of the brightest features in the aurorae is something called the 'Io footprint'. Charged particles from Jupiter's tiny moon Io are caught in the Jovian magnetosphere and

(a)

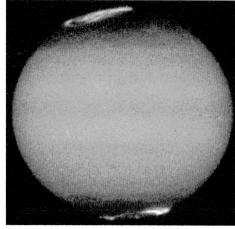

(b)

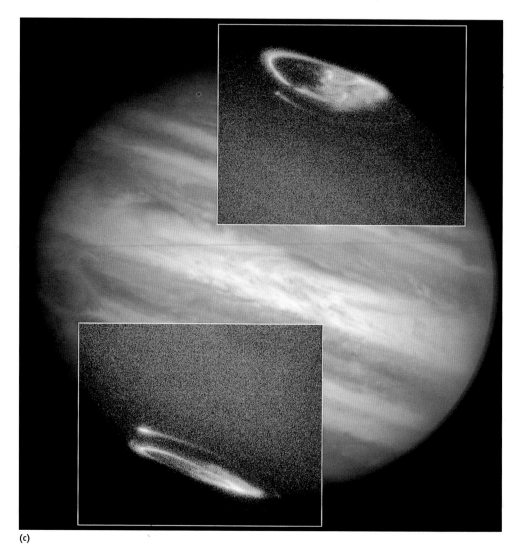

(c)

Figure 2.11. The Jovian aurorae as seen by HST's WF/PC-2 and STIS instruments. The false-color images in **(a)** and **(b)** reveal how the magnetic field is offset from Jupiter's spin axis by 10–15 degrees. In **(a)** the north auroral emission is rising while the south auroral oval is setting. In **(b)** (taken at a different time) the positions of the ovals have shifted. **(c)** shows a STIS image of the ovals. The small streak equatorward of each oval is the electromagnetic 'footprint' of Io as it orbits Jupiter. *(John T. Clarke, University of Michigan; John Trauger, Jet Propulsion Laboratory; STScI; NASA; ESA)*

drawn to the poles in a 'flux tube'. The 'footprint' of the flux tube rotates across Jupiter as Io orbits the planet. Clarke and his colleagues measured the magnetic footprint of Io to be about 1000 to 2000 kilometers across, speeding across the planet from east to west at more than 5 kilometers per second. If an observer could somehow be transported to Jupiter to see this phenomenon, it would be an experience unlike any auroral storm seen on Earth. 'If you were at Jupiter's cloud tops, under the Io footprint,' says Clarke, 'the aurora would fill the entire sky. You would see an explosion as the gases 250 miles over your head heated to more than 10 000 degrees Fahrenheit.'

Understandably, the data collected about about the Jovian aurorae and magnetic field were of great interest to the Galileo science teams as their spacecraft began its long Jovian mission and repeated trips through the powerful magnetosphere of the planet. Clarke's team shared their auroral and magnetic field data with the Galileo team; and eventually combined Hubble and Galileo information will help these researchers come up with a more accurate understanding of the exact source on Io for the charged particles that stream into the Jovian magnetosphere every second.

Because Io is the Solar System's most dynamic moon, it has been the impetus for a series of HST studies. These include a continuing series of ultraviolet atmospheric observations, surface mapping, and studies of the complex interactions between Io and Jupiter. This rocky, silicate-rich moon is under continual study by both professional and amateur astronomers, using ground-based facilities along with HST. Before Voyager, scientists suspected the presence of some kind of volcanic activity on Io, and, indeed, both spacecraft showed Io to be an incredibly active world. At least eight eruptions were seen by Voyager 1, and Voyager 2 imaged six volcanoes. Squeezed into an orbital resonance between Jupiter, Ganymede, and Callisto, Io is caught in strong gravitational fields, leading to flexing, which causes heating and melting. Eventually, the 'melt' (molten interior) reaches the surface in the form of sulfur flows and plumes which send fountains of sulfur dioxide up through Io's tenuous atmosphere.

Some of the first Faint Object Camera images of Io's surface were taken by Francesco Paresce of the Space Telescope Science Institute and Paola Sartoretti of the University of Padova, Italy. Comparing these images with those from Voyager shows dramatically that the moon really has not changed very much in the years since the Voyager encounter, despite continual volcanic activity.

In Figure 2.12, the upper right image is an ultraviolet scan. Notice that it looks very different from the visible light image on the left. This is probably because Io is covered with sulfur dioxide frost, which absorbs ultraviolet light and causes those areas to look dark. The same areas look bright in visible light.

Io orbits Jupiter once every Earth day, and for half the orbit it is hidden behind Jupiter, out of sunlight. Astronomers have noticed that sometimes Io looks about 10 to 15 per cent brighter when it emerges from Jupiter's shadow. This phenomenon is called Io post-eclipse brightening and was the subject of early HST observations by amateur astronomer Jim Secosky. Secosky, a high-school biology teacher, said he ran across some material on Io's brightening when he was researching a paper on water in the Solar System. He wanted to

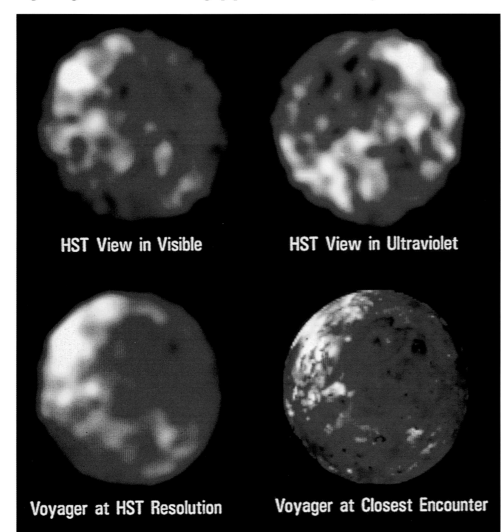

Figure 2.12. FOC and Voyager images of Io. *(Francesco Paresce, ESA and STScI; Paola Sartoretti, University of Padova; NASA)*

Figure 2.13. (a) (Opposite) An image of Jupiter taken on May 18, 1994, by WF/PC-2. Jupiter was 670 million kilometers from Earth at the time the image was taken. The Jovian moon Io floats above the planet, casting its shadow on Jupiter (dark spot). Even from this distance, HST was able to resolve some surface details of the tiny moon. *(H.A. Weaver and T.E. Smith, STScI; J.T. Trauger and R.W. Evans, NASA–Jet Propulsion Laboratory; NASA)* (b) (left) HST snapped this ultraviolet image of Jupiter and Io as part of a series taken by the WF/PC-2 in concurrent study with the Galileo spacecraft. *(J. Spencer, Lowell Observatory; NASA)*

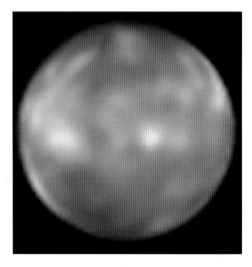

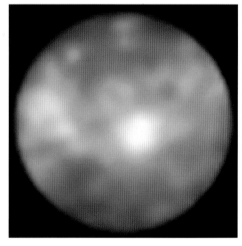

Figure 2.14. HST images of Io, taken in March 1994 **(upper)** and July 1995 **(lower)**, record changes on the surface of this volcanic moon. *(J. Spencer, Lowell Observatory; NASA)*

track down a cause for the brightening and to determine whether evaporation of sulfur dioxide frost was a possible explanation. His HST program first ran in the spring of 1992, but his results were ambiguous, possibly due to timing problems with the observation. He was not able to observe Io exactly after eclipse, although he had three snapshots of the limb of Jupiter to try to spot the phenomenon. Still, Io does seem to brighten from time to time as it comes out of Jupiter's shadow, and Secosky concluded that the brightening could be driven by sporadic volcanic activity.

Early in HST's observations of Io, various teams of observers turned HST's spectrographic attention to the ultraviolet emissions of sulfur dioxide from Io. The Faint Object Spectrograph helped confirm a variable sulfur dioxide atmosphere on Io, as a result of sulfur and oxygen emissions from Io's surface. Io's atmosphere is three times smaller than previously thought and shows an interesting tendency to be thicker over regions where sulfur ice is 'melting' or where volcanic vents are pumping sulfur from underground. After Voyager passed by in 1979, planetary scientists theorized that Io's atmosphere was a result of emissions from volcanoes and evaporation of surface frost in sunlit areas. They also established an upper limit for the extent of the atmosphere at 5 Io diameters. Now it is thought to be an average of 1.5 Io diameters, thicker over known volcanic regions and frost deposits and thinner in other places.

In July 1996, HST caught Io's volcano Pele in the act of eruption. The telescope snapped a picture of a 400-kilometer-high plume of gas and dust erupting from Io's largest volcanic vent. Until this observation, only near-Jupiter spacecraft had glimpsed volcanic plumes on Io. Now HST can keep a continual eye on the little moon, watching out for more volcanic action.

Some of the sulfur in the volcanic plumes falls back onto Io's surface as a kind of 'snow', while some of it feeds into a giant ring of high-temperature gas surrounding Jupiter called the Io plasma torus. As discussed above, ionized sulfur in the torus is one source of the particles that form the Jovian aurorae. During HST's observations, however, oxygen emissions from the torus itself were discovered. It turns out that oxygen is twice as abundant in the torus as sulfur. Recent observations of the torus using the Space Telescope Imaging Spectrograph have led researchers to map the actual shape of the torus, and to search for the energy source which causes half of the ring to glow more brightly than the other half.

Io is not the only Jovian moon studied by HST. Europa, the icy little moon celebrated in Arthur Clarke's science fiction books *2010* and *2061*, recently came under HST's unblinking gaze, as did its sister worlds Ganymede and Callisto. The result of the Europa campaign was the discovery of a tenuous atmosphere of molecular oxygen around this little world that is approximately the size of Earth's moon. On Earth, the presence of oxygen indicates the presence of life, but on Europa, the discovery of oxygen points to a non-biological process. In this case, the moon's icy surface is exposed to sunlight. In addition, the Europan terrain is impacted by dust and charged particles trapped in the Jovian magnetic field.

Heating from these processes is just enough to cause the frozen water ice to melt, thus producing water vapor and gaseous fragments of water molecules. A series of chemical reactions ultimately produce hydrogen (which escapes to space) and oxygen, which manages to accumlate and form an atmosphere extending about 200 kilometers above the Europan surface. Coupled with in-situ observations of Europa's cracked, icy surface by the Galileo spacecraft, scientists will use HST observations to try to understand the interior structure of this world – and to answer questions about whether it could harbor and support life.

Figure 2.15. HST caught one of Io's most powerful volcanoes, named Pele after the Hawaiian goddess of the volcano, in mid-eruption off the lower left limb. This was the first time one of Pele's volcanic plumes had been imaged since the Voyager 1 fly-by in 1979. *(STScI; NASA)*

Figure 2.16. (Bottom left) HST image of Jupiter's water moon Europa on October 9, 1995. *(J. Spencer, Lowell Observatory; K. Noll, STScI; NASA)*

Figure 2.17. (Bottom center) HST image of Ganymede, taken on October 9, 1995, Jupiter's largest moon and the largest satellite in the Solar System. *(J. Spencer, Lowell Observatory; K. Noll, STScI; NASA)*

Figure 2.18. (Bottom right) HST image of Callisto, taken on October 9, 1995. *(J. Spencer, Lowell Observatory; K. Noll, STScI; NASA)*

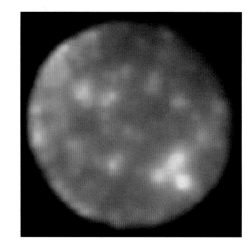

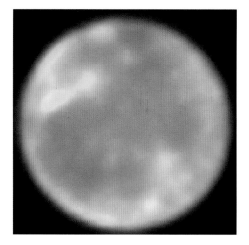

HST mapped major surface features on Callisto and Ganymede. From Earth orbit, impact craters and albedo features of some kind are clearly visible in HST's images of these tiny worlds. In addition, Callisto appears to have deposits of sulfur dioxide in some form on its leading hemisphere, and the FOS saw evidence of it . Presumably these deposits result from collisions with charged particles in the Jovian magnetosphere. Evidence of fresh ice on Callisto's surface indicates that this little moon continues to be bombarded with micro-meteorites that 'garden' its surface.

Saturn

The gas giant Saturn, with its glittering ring system, has long been a favorite of both amateur and professional astronomers. So it should come as no surprise that Saturn was selected as one of the first planetary objects to be studied by HST. In fact, Saturn presented HST with its first planetary challenge only four months after launch.

On August 26, 1990, HST took what many called a 'Voyager quality' image of the planet. Figure 2.19 shows Saturn as it would appear from the ground if it were only twice as far away from us as the Moon. In reality, Saturn is 1.4×10^9 kilometers away – or about 3600 times farther away than the Moon. HST imaged the northern hemisphere of Saturn, showing the banded structure in the planet's upper cloud decks. The famous 'Cassini Division' (the largest dark gap) in the rings is clearly visible, as is a thin division near the edge of the ring system. This is the Encke Division, which has never been photographed from the Earth. The faint innermost 'crepe ring' is also shown.

About one month after HST's first image of Saturn was taken, a pair of amateur observers noticed a large white spot developing in Saturn's northern equatorial region. The date was September 25, 1990. They announced their find, and shortly afterward Reta Beebe did a series of ground-based observations. She reported that the storm was expanding eastward at about 400 meters per second, smearing out across the planet's cloud tops. Immediately, the HST

Figure 2.19. WF/PC-1 view of Saturn reconstructed from red, green, and blue light images taken on 26 August, 1990. Image deconvolution was applied. *(NASA)*

observing community clamored for time on the telescope to observe this 'target of opportunity'. Everyone knew the telescope was still in orbital verification mode, and that procedures to track a moving target were barely functional, but Beebe, along with members of the Wide Field and Planetary Camera team, convinced the Institute Director (Riccardo Giacconi) that this was an event worthy of special attention.

On November 9, 1990, after a hurried few weeks of planning, HST was aimed at Saturn, and took a series of time-lapse images. The storm turned out to be so dynamic and complex that observers pleaded for more and more time to catch every nuance before it faded away. Life became very interesting very quickly for the observers and the moving targets planning group; they had to rush to put together a second observing run even before the first was completed. A request for a second observation came in only a week before Saturn would be too close to the Sun to be observed with HST. If the planners missed the opportunity, it would be 100 days before Saturn emerged from behind the Sun, and the storm might be gone. The decision was made to proceed, and everybody had just over a day to deliver the observation parameters and spacecraft commands to Goddard for execution.

Two problems threatened to complicate matters for the Saturn observers. First, the telescope gathered so much data that there was some danger of overwriting the data buffers on the spacecraft. Secondly, it also happened that the US military was running a secret shuttle mission during the week of the Saturn observations. This put a heavy load on the Tracking

Figure 2.20. HST color-enhanced image of Saturn's Great White Spot, taken a month after the spot's discovery. The turbulent upper atmosphere had already smeared the spot's clouds across the planet. *(STScI; NASA)*

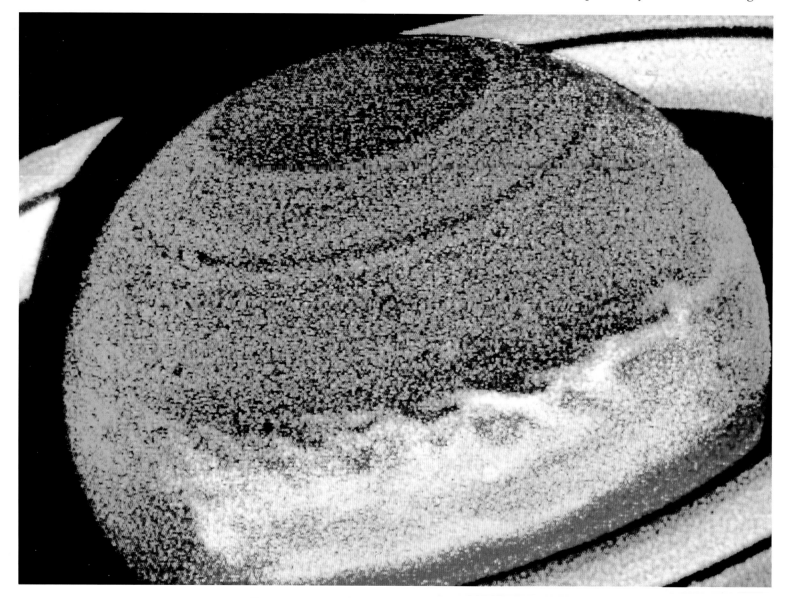

Data and Relay Satellite System, which, in turn, threatened the flow of HST data. Fortunately, controllers managed to work out the difficulties.

Although the storm in the upper cloud belts came as a surprise to many people, it is actually a type of event that occurs relatively frequently – about once a generation. Planetary scientists first characterized it as Saturn's answer to Jupiter's Red Spot, but there are major differences between the two storms. On Saturn, high winds make such large-scale atmospheric events very short-lived phenomena. The event turned out to be less like a storm and more like an eruption from deep below the Saturnian cloud decks. One scientist was quoted in *Science* as saying, 'Saturn burped'. Indeed, the 'Great White Spot of 1990' has now been explained as an upwelling plume of ammonia ice crystals that burst out through the cloud tops. To see it so shortly after HST's launch and deployment was truly serendipitous.

Figure 2.21. This image of Saturn, taken on December 1, 1994, shows a rare storm that appeared in September 1994. The arrowhead-shaped storm was generated by an upwelling of warmer gas from deep within the lower clouds. *(R. Beebe, New Mexico State University; D. Gilmore, L. Bergeron, STScI; NASA)*

Figure 2.22. This HST image of Saturn was taken in May 1995, shortly before the ring-plane crossing. The rings did not disappear completely because the edge of the rings reflected sunlight. The dark band across the middle of Saturn is the shadow of the rings cast on the planet. The bright stripe directly above the ring shadow is caused by sunlight reflected off the rings onto Saturn's atmosphere. The tiny starlike objects in the ring plane were two of Saturn's icy moons. They are Tethys (left, slightly above the ring plane) and Dione (right). Though Saturn was approximately 1440 million kilometers away, Hubble saw details as small as 725 kilometers across. *(Amanda S. Bosh, Lowell Observatory; Andrew S. Rivkin, University of Arizona/ Lunar and Planetary Laboratory; the HST High Speed Photometer Instrument Team; R.C. Bless, University of Wisconsin; NASA)*

HST observations of Saturn continue in an effort to gauge the long-term weather patterns in the clouds. In late 1994, Reta Beebe and Space Telescope Science Institute's Glenn Bergen imaged Saturn and spotted yet another storm. Like the 1990 storm, this one was primarily a white cloud of ammonia ice crystals that formed when warm gases flowed from underneath the colder, upper cloud decks. It now seems that these storms occur in a cycle, usually appearing during northern hemisphere summer. Saturn's high-speed winds, clocked up to 500 meters per second, are spreading out the storm of 1994. As with the 1990 white spot, HST continued to track the storm's progress throughout the following months.

Few sights in the Solar System compare to the beauty of Saturn's rings. They have been known to astronomers since Galileo's time. They so puzzled him at first that he was at a loss for words to describe these 'handles' on Saturn. Indeed, the true nature of the rings kept astronomers conjecturing for years. Were they solid or made up of particles? Did they rotate? If so, how fast? If they were solid, how did they manage to stay together as they rotated?

Today we know that the rings are vast collections of particles, stretching some 200 000 kilometers across, but only a few kilometers thick. Thanks to the Voyager 2 spacecraft, we know that they sport an amazingly complex structure. During the August 1981 encounter, Voyager observed a star move behind the rings and recorded light intensity fluctuations as the starlight passed through. The study of those fluctuations (called 'occultation work'), coupled with some very detailed imagery, revealed an amazing array of complex phenomena in the rings: gaps, moonlets, spiral density waves, and narrow ringlets. However, Voyager was only one spacecraft, and the next spacecraft to go out Saturn's way – the Cassini probe – did not leave Earth until late 1997 and is due to arrive in 2004. Until then, scientists need a way to carry out occultation work from Earth.

For an occultation to be successful, the telescope, the rings, and the star they will occult have to be aligned perfectly. Using HST before the servicing mission, such things were possible, but incredibly difficult. The High Speed Photometer was perfectly designed to do high-resolution occultations from Earth orbit, but, of course, it was limited by the spherical aberration and jitter problems. The problems were daunting: the position of the planet had to be known precisely; the location of the star also had to be quite accurately known

Figure 2.23. Saturn's magnificent ring system seen tilted edge-on in this HST WC/PC-2 image taken on August 10, 1995, when the planet was 895 million miles (1 440 million kilometers) away. Several of Saturn's icy moons are visible as tiny starlike objects in or near the ring plane. They are, from left to right, Enceladus, Tethys, Dione, and Mimas. *(Phil Nicholson, Cornell University; NASA)*

Figure 2.24. Rings of Saturn and the HSP. The path of a star behind the rings during an occultation is shown. The dots indicate 5-minute intervals during an observation period of 20 hours. Data were taken in the intervals marked by the green dots and were not taken due to the South Atlantic Anomaly (red dots) or due to Earth occultations (violet dots). Analysis of the data revealed occultations by 43 different ring features (see page 26). *(Robert Bless, University of Wisconsin and the HSP Team)*

so that HST could follow it; and the position and stability of the telescope itself was important. Despite the complexity of the task, the HSP team went ahead with an occultation in October 1991 (see also Figure 1.21, page 26).

Figure 2.24 shows the path of the occulted star behind the rings. The most striking result from this experiment was that 43 different ring features could be identified using the HSP's data, despite problems with spherical aberration and jitter. It is significant that the observation worked as well as it did, and even though the HSP was removed from HST in 1993, the work proves that successful occultations can be done from Earth-orbiting spacecraft.

Saturn's rings were again the focus of attention in 1995 due to a phenomenon that occurs every 15 years called 'ring-plane' crossing. As Saturn and Earth orbit the Sun, there are periods of time when the 100-kilometers thick rings are seen edge-on. In addition to measuring the starlight shining through the rings, planetary scientists wanted to use HST to measure the thickness of the rings, study the rings' atmosphere, look for the known satellites of Saturn within the rings, and – if possible – to find satellites that Voyagers 1 and 2 may not have seen.

One of the most surprising finds was the existence of new moons orbiting Saturn. Images taken on May 22, 1995 showed four new moons, orbiting near Saturn's eccentric 'F' ring. This discovery follows in a long tradition of moon discoveries during ring-plane crossings. Thirteen of Saturn's moons were discovered during crossings ranging between the years 1655 and 1980. All the others were discovered by the Voyager spacecraft during the two fly-by missions in the early 1980s. Saturn's aurorae and atmosphere continue to pique research interest. STIS has studied the aurorae, focusing in on emissions from hydrogen in two states. Unlike Earth's aurorae, Saturn's electromagnetic storms glow only in ultraviolet light. By contrast, in the infrared, the hazes and clouds that populate Saturn's upper atmosphere were a prime target for the NICMOS instrument. In the infrared, the structure of the atmosphere and rings takes on a different look.

Before leaving the Saturnian system, we should make a stop at Titan – the largest, and one of the more intriguing, moons orbiting Saturn. Voyager showed it to be a world shrouded with methane clouds. Photochemical smogs float high in its atmosphere, which consists primarily of nitrogen, with a small amount of methane. Its present temperature, less than

Figure 2.25. Saturn in ultraviolet and infrared. **(a)** HST's STIS instrument reveals details in Saturn's aurorae that only appear at ultraviolet wavelengths. Ripples, patterns and brightenings indicate that the aurora is shaped and powered by a continual tug-of-war between Saturn's magnetic field and the solar wind. *(J.T. Trauger, Jet Propulsion Laboratory; NASA)* **(b)** An infrared view of Saturn, taken by NICMOS in early 1998, presents the ringed planet as a study in pastels. These green and yellow colors indicate a haze above the main cloud layer. Red and orange indicate clouds reaching high into the atmosphere, The satellites Dione and Tethys appear in the lower left and upper right, respectively. *(Erich Karkoschka, University of Arizona; NASA)*

(a)

(b)

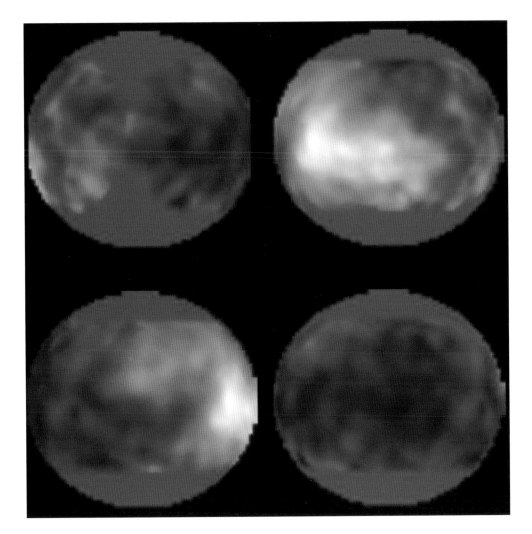

Figure 2.26. These four global projections of Saturn's largest moon Titan were assembled from 14 images taken by WF/PC-2 between October 4 and 18, 1994. The upper left image is the hemisphere which faces Saturn. The upper right image is the leading hemisphere and shows the brightest region. The lower left image is the hemisphere which faces away from Saturn, and the lower right image is the trailing hemisphere. *(Peter H. Smith, University of Arizona/Lunar and Planetary Laboratory; NASA)*

−178 degrees Celsius, keeps methane and ethane in gaseous and liquid states, but is cold enough to make water ice as hard as rock.

The thick clouds and haze prevented Voyager from seeing Titan's surface, and scientists had to wait until HST's Wide Field and Planetary Camera-2 was installed to take a near-infrared look under the clouds. Led by University of Arizona's Peter Smith and Mark Lemmon, a group of HST observers looked at Titan during a complete 16-day rotation of the moon. They took advantage of the fact that Titan's smog layer is transparent to near-infrared light and mapped surface features according to how much light they reflected. The image resolution of this technique allowed the researchers to see objects as small as 576 kilometers across.

One large bright area appeared to be just over 4000 kilometers wide (about the size of Australia) and represents some sort of solid surface area. Other bright and dark regions could be continents, oceans, craters, or different kinds of surface features. Continuing analysis of the HST data will yield new information on small-scale surface features and atmospheric winds. HST's Titan information will be useful to the planners of the Cassini mission to Saturn, which will release the probe Huygens to study the atmosphere and surface of Saturn's cloud-shrouded largest moon.

Uranus and Neptune

In 1986, Voyager 2 flew past the gas giant planet Uranus, and gave planetary scientists their first close-up look at this enigmatic place. It appeared as a featureless blue ball, with a mainly hydrogen atmosphere. Voyager's ultraviolet cameras revealed cloud features and

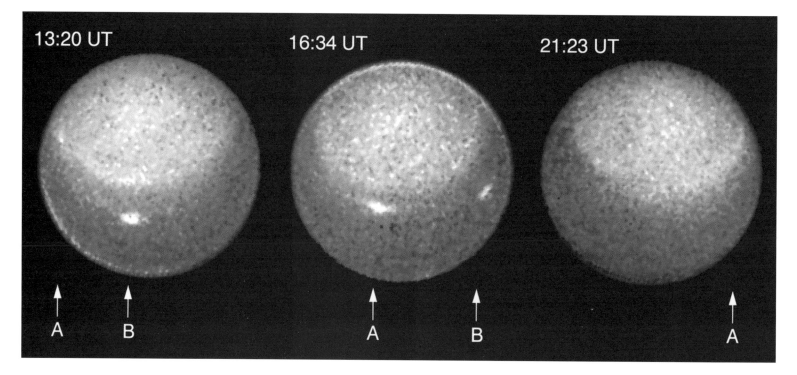

13:20 UT 16:34 UT 21:23 UT

A B A B A

Figure 2.27. Images of Uranus taken by WF/PC-2 on August 14, 1994, showing the planet's rotation and high-altitude clouds (marked A and B). Uranus rotates once every 7 hours, 14 minutes. These atmospheric details were previously seen only by the Voyager 2 spacecraft in 1986. (*Kenneth Seidelmann, US Naval Observatory; NASA*)

Figure 2.28. HST false-color image of Uranus, using infrared filters to capture detailed features of three layers in the Uranian atmosphere. The red around the planet's edge represents a very thin high-altitude haze. The yellow near the bottom is another hazy layer, and the deepest layer (in blue near the top of Uranus) shows a clearer atmosphere. Image processing was also used to bring out the rings, which are actually as dark as black lava or charcoal. (*Erich Karkoschka, University of Arizona Lunar and Planetary Laboratory; NASA*)

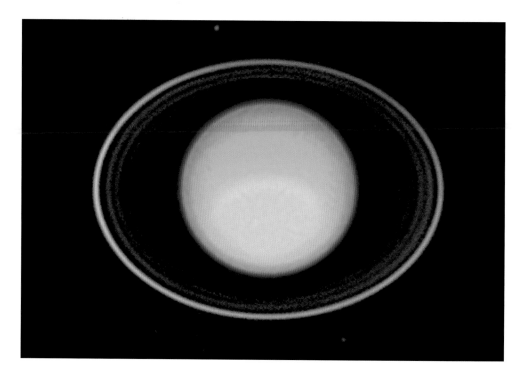

high-altitude methane hazes floating over the planet's sunward-facing southern pole. Voyager also imaged several of the moons of Uranus. These little worlds showed unusual variation in their surface features, ranging from the darkened ice of Umbriel to the wildly variable terrain of Miranda.

Eight years after Voyager 2's fly-by, a team of astronomers led by Ken Seidelman of the US Naval Observatory began a program of Uranus observations with HST. In their first image, taken on August 14, 1994, clouds, rings, and moons appear in sharp detail. The on-going observation program will focus on determining the composition and precise orbits of the moons and rings, and on detecting seasonal changes in the high-altitude hazes that blanket the planet.

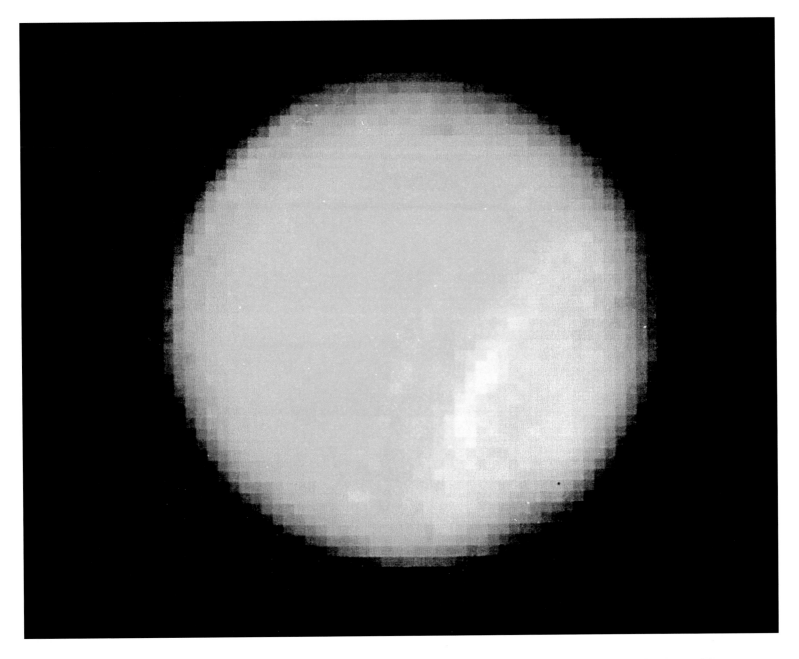

Neptune is the farthest gas giant planet observed by HST, and it appears to have changed in appearance since Voyager 2 visited in 1989. Similar to Uranus in atmospheric composition (hydrogen, helium, and methane), Neptune nonetheless distinguished itself by sporting a set of storms that were quickly dubbed 'the Great Dark Spot' and 'Dark Spot 2' by Voyager scientists. These storms raced across the planet's disk, revealing wind speeds as high as 325 meters per second. Along with the storms, a series of high-altitude, cirrus-like clouds set Neptune apart from its distant neighbor.

Two groups of scientists have zeroed in on Neptune, using the WP/PC-2 to obtain high-resolution images of the planet. A team led by John Trauger, Principal Investigator for WF/PC-2, and including Jet Propulsion Laboratory astronomer David Crisp and Massachusetts Institute of Technology astronomer Heidi Hammel, acquired the first high-resolution images in June 1994. Because HST can resolve the disk of Neptune at least as well as ground-based instruments can resolve Jupiter, it is in an ideal position to monitor atmospheric changes on the much more distant planet. Ultraviolet images show some of the same features in Neptune's upper atmosphere that Voyager saw. The bright clouds are probably well above the main cloud deck and the methane layer that causes Neptune to look so blue. According to Heidi Hammel,

Figure 2.29. True-color picture of Neptune constructed from red, green, and blue WF/PC-2 images. The blue color of the planet stands out. *(David Crisp and John Trauger, NASA–Jet Propulsion Laboratory and the WF/PC-2 Science Team; H. Hammel, MIT; NASA)*

Figure 2.30. (a) (Opposite top) WF/PC-2 images of Neptune (first hemisphere) in different wavelength bands, taken on June 28-29, 1994. The wavelength in angstroms is given by ten times the number shown and the filter is either wideband (W) or narrow-band (N). The images marked CH4 isolate light from methane absorption bands. The bright clouds are most obvious in these filters. **(b) (Opposite bottom)** Same as in (a) but Neptune has rotated by about 180 degrees and the second hemisphere is shown. The Great Dark Spot, which covered almost 40 degrees of longitude at the time of the Voyager 2 fly-by, has disappeared. *(David Crisp and John Trauger, NASA–Jet Propulsion Laboratory and the WF/PC-2 Science Team; H. Hammel, MIT; NASA)*

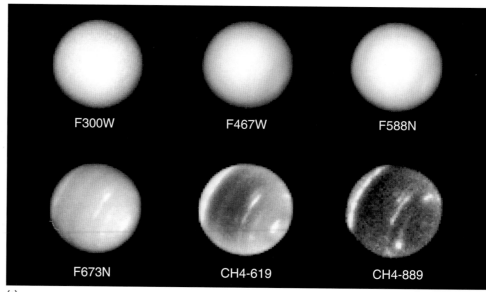

(a)

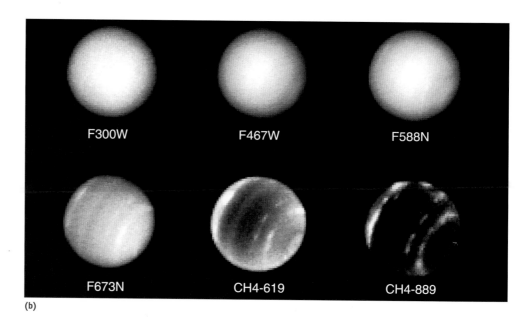

(b)

Figure 2.31. (Below) New clouds on Neptune, imaged on April 19, 1995, by HST. *(H. Hammel, MIT; NASA)*

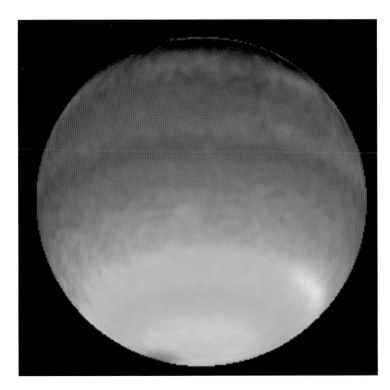
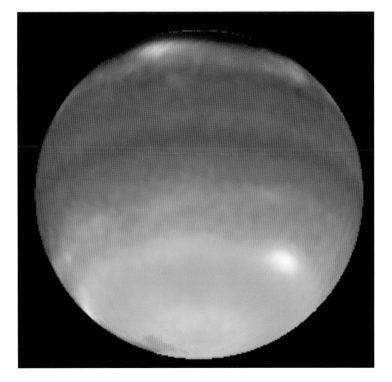

the biggest discovery in the HST data is that the Great Dark Spot and Dark Spot 2 have disappeared. The disappearance is not completely understood, but it does emphasize the changing nature of the Neptunian atmosphere.

A more comprehensive program of Neptune studies was executed late in 1994 by Heidi Hammel and Wesley Lockwood of Lowell Observatory. Surprisingly enough, in the images from that program, Neptune was sporting a new dark spot in its northern hemisphere. This spot is a mirror image of the Great Dark Spot first mapped by Voyager 2 in 1989, and, according to Hammel, the latest HST image shows that Neptune has changed radically since the Voyager visit. 'New features like this indicate that with Neptune's extraordinary dynamics, the planet can look completely different in just a few weeks,' she explained.

Pluto

Pluto is the ninth planet, and so far it is the only one that has not been visited by a spacecraft. It is considered a 'weird' little place, and – with its companion Charon – is often referred to as a double planet.

Pluto probably deserves its deviant, peculiar reputation. It orbits 5.9×10^9 kilometers from the Sun in a tilted, highly elliptical orbit. Currently, it is inside the orbit of Neptune, temporarily making it the eighth planet out from the Sun. Pluto's 'year' is 249 Earth-years long, so only a small fraction of one Pluto 'year' has passed since Clyde Tombaugh discovered it in 1930.

HST's contributions to Pluto studies fall in the realms of surface mapping, and imaging of Pluto and Charon together. HST imaged most of the surface of this distant, icy world in mid-1994. Variations in bright and dark patterns across Pluto's surface could be caused by impact craters or basins. However, it is quite likely that the planet's thin nitrogen–methane atmosphere is freezing out and depositing complex patterns of icy frosts across the surface of Pluto. As the planet heads into its 60-year-long winter, those patterns could intensify as more of the atmosphere condenses as frost.

Because Pluto is so distant, nobody knew that it had a companion until 1978, when Charon was discovered by James Christy at the US Naval Observatory. The two worlds orbit each other at a distance of 19 000 kilometers, and, from our viewpoint on Earth, it has been nearly impossible to tell the two apart, much less determine the two worlds' orbital periods around

Figure 2.32. (Above) These images are part of a series made by HST during nine orbits spanning one 16.11-hour rotation of Neptune on August 13, 1996. The planet's blustery weather was the object of study in this series. Clouds elevated above most of the methane absorption (as seen by HST) appear white; the very highest clouds are a yellow-red color. The dark blue belt just south of the equator is the zone where winds blow at nearly 1450 kilometers per hour. The green belt indicates a region where the atmosphere absorbs blue light. *(Lawrence Sromovsky, University of Wisconsin-Madison; NASA)*

(a)

(b)

(c)

each other. It was not until HST looked at the pair in 1990 that astronomers were able to clearly separate both components of this 'double planet'. Figure 2.33a is the best ground-based image of the pair, taken from the Canada–France–Hawaii Telescope atop Mauna Kea. 2.33b shows the first HST image of the pair, with an angular separation of 0.9 arcseconds. With precise measurements of the orbits estimated by using HST data, astronomers are refining their estimates of the masses and densities of these two worlds. Coupled with ground-based studies of the pair, astronomers think Pluto is about 60 per cent rock, whereas Charon comprises mostly ice.

In 1994, HST observed Pluto and Charon again as seen in figure 2.33c. In the new image, the pair appear as disks. At least one set of Pluto/Charon observations shows some striking differences between the two. For one thing, the HST images have helped refine the two worlds' respective sizes: Pluto is 1440 kilometers across and looks to have smooth ice covering its rocky interior. Charon, on the other hand, looks bluer than Pluto and is only about 790 kilometers across, about half the diameter of Pluto. Pointing to these differences, some astronomers question whether the pair evolved together or have been thrown together by cosmic circumstance when the Solar System was forming.

It is possible that the pair was created, along with other similar 'planetary embryos', in the outer fringes of the primordial solar nebula. The other embryos were either used up in the creation of the gas giant planets, or expelled out of the cloud. Pluto and Charon may be the only large leftovers. Since they are more unalike than alike, possible future missions such as the Pluto Fast Flyby may reveal more about their origins than HST's observations can. Until then, continuing observations of Pluto with HST will supply mission planners with a wealth of information as they decide how best to observe Pluto and Charon up close.

Figure 2.33. (Opposite) (a) The best ground-based image of Pluto and Charon. **(b)** FOC image (before COSTAR) of Pluto and Charon. The two were separated by 0.9 arcsecond. **(c)** This FOC image (post-COSTAR) shows these bodies as distinct and sharp disks. *(R. Albrecht, ESA/ESO Space Telescope European Coordinating Facility; NASA)*

Figure 2.34. (Above) Hubble imaged nearly the entire surface of Pluto as it rotated through its 6.4-day period in March 1996. These blue-light computer reconstructions show that this distant planet is an unusually complex object, with more large-scale surface contrast than any planet except Earth. Some of the variations across Pluto's surface may be caused by basins or fresh impact craters. Most of the features imaged by Hubble are likely due to the distribution patterns of frost that migrate seasonally across the planet's surface. *(A. Stern, Southwest Research Institute; Marc Buie, Lowell Observatory; NASA; ESA)*

HST and comets

Comets are a perennial favorite of skywatchers. These icy bodies spend much of their time locked in the deep-freezers of the outer Solar System – the Kuiper Belt and Oort Cloud. Because they formed at the same time as the Solar System, they are also tracers of our past. They contain chemical mixtures prevalent when the planets were forming, some 4.5 billion years ago. It is only when comets are tugged from their orbits in the outer Solar System into the relatively warmer climes of the inner Solar System that they are heated. They sublimate

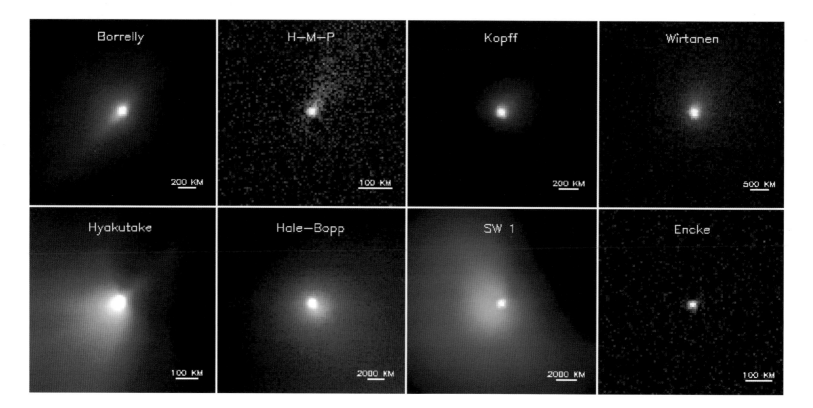

Borrelly | H–M–P | Kopff | Wirtanen

200 KM | 100 KM | 200 KM | 500 KM

Hyakutake | Hale–Bopp | SW 1 | Encke

100 KM | 2000 KM | 2000 KM | 100 KM

Figure 2.35. This is a series of Hubble images of all the comets (except Shoemaker-Levy 9) observed with the telescope since the First Servicing Mission in 1993. The comets are: Borrelly, Honda-Mrkos-Padjusakova (H-M-P), Kopff, Wirtanen, Hyakutake, Hale-Bopp, Schwassman-Wachmann 1 (SW 1), and Lncke. This 'family portrait' shows the diversity in morphology of the inner comae of these comets. Their diameters have been estimated from these images: Borrelly is 7.8 kilometers; H-M-P is 0.44–1.68 kilometers; Kopff is 3.6 kilometers; Wirtanen is 1.2 kilometers; Hyakutake is smaller than 15 kilometers; Hale-Bopp is 30–40 kilometers; SW 1 is 30–40 kilometers; Encke is 0.9 kilometers. *(H. Weaver, Johns Hopkins University; NASA)*

and, as they do, they give off the water vapor, carbon dioxide, carbon monoxide, and many other gases that provide clues to their makeup.

Since they are rapidly moving objects, comets present special challenges to the tracking capability of HST, but the scientific returns can be magnificent. HST has had the good fortune to observe more than a dozen comets since its launch. They can be imaged with the Wide Field Camera 2, and they have been studied in ultraviolet light by the spectrographs to determine their chemical compositions, the structures of their comae, and the activity of their dust and plasma tails.

The first comet to be studied by HST was Comet Levy 1990c. Space Telescope Science Institute scientist Hal Weaver, in collaboration with University of Maryland astronomer Michael A' Hearn and Institut d'Astrophysique of Belgium astronomer Claude Arpigny, were particularly interested in an arc of dust that had been spotted in the tail by a ground-based observatory. They used HST's Wide Field Camera, in conjunction with the European Southern Observatory in Chile, to search for that arc, and came up with a series of images showing the dust.

Astronomers used HST's spectrographs to compile an inventory of the chemicals that make up comets. For example, Arpigny and Weaver worked on a series of observations that looked at 'Cameron band' emissions in carbon monoxide from Comet Hartley 2. These emissions are probably produced when sunlight photodissociates (breaks down) carbon dioxide to carbon monoxide. Measuring these emissions is an important indicator of how much carbon dioxide exists in a comet.

The middle years of the 1990s produced a fine array of comets that HST could study. Thirteen comets made their way across HST's gaze, allowing comet researchers to study what they call the 'morphology' or physical appearance of the comets. Although HST could not peer through the coma structure to see the icy nucleus of each comet, the telescope was extremely helpful in giving scientists insight into the structure of the coma of each comet, the dust production of each comet as its ices melted during close approach to the Sun, as well as the makeup of the gases in cometary plasma tails. In addition to comets like Hyakutake, Hale-Bopp, and Wirtanen, HST also imaged icy Chiron, which is thought to be a large cometary body that comes no closer to the Sun than the orbit of Uranus. Probably no comets were stud-

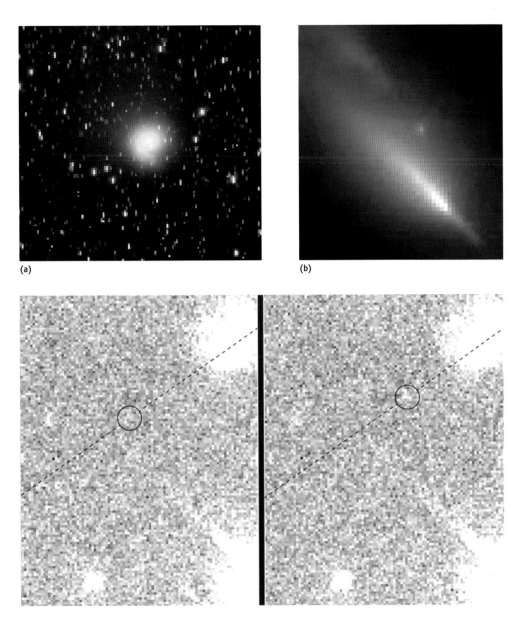

Figure 2.36. (a) HST's view of Comet Hale-Bopp on October 5, 1995. The peculiar spiral structure of the jet was due to the rotation of the comet's nucleus. *(H. Weaver, Applied Research Corporation; P. Feldman, Johns Hopkins University; NASA)* **(b)** HST 'close-up' of Comet Hyakutake's nucleus and clouds that came off the nucleus on March 24, 1997. At least three separate clouds came off and appeared to accelerate down the tail of the comet. *(STScI; NASA)*

Figure 2.37. HST's first image of a Kuiper Belt object – a cometary nucleus. The pair of images shows the apparent motion of the object. *(A. Cochran, University of Texas; NASA)*

ied as much as Shoemaker-Levy 9 (discussed earlier) and Hale-Bopp, which graced our skies through the first half of 1997. HST was used to image Hale-Bopp until it was too close to the Sun for the telescope. During its observations, HST scientists paid particular attention to the jets emanating from the comet, as well as the ever-changing coma.

Cometary nuclei are the latest bits of the outer system to come under the HST's high-resolution gaze. Out beyond the orbit of Neptune lies a large reservoir of cometary ice chunks called the Kuiper Belt. This deep-freeze at the fringe of the Solar System has long been suspected to be the 'starting point' for a number of comets that have been observed from Earth. The problem has been that these objects – called Kuiper Belt Objects, are so small and dark that they evade detection. Scientists working on the problem of determining how populous the Kuiper Belt is have largely had to rely upon indirect arguments to make a case for the existence of the collection of icy objects. In 1994, HST took a series of 5-hour exposures looking for the proverbial cometary 'needle in the haystack' millions of kilometers from Earth. A slow moving object appeared in several of the images, and that object is generally thought to be the first evidence for the existence of Comet Halley-sized nuclei in the Kuiper Belt. Since 1994, astronomers have identified more than 40 large-sized objects – so-called proto-comets – in this region of interplanetary space. The discovery of a 'general-case' type proto-comet using HST gives new impetus to the search for cometary nuclei using the telescope.

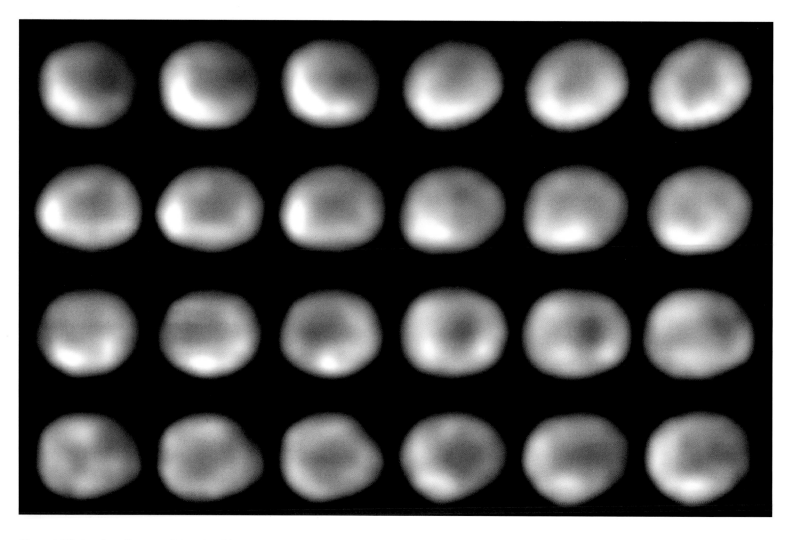

Figure 2.38. A series of images of the asteroid Vesta, taken with HST on April 19, 1995. *(Ben Zellner, Georgia Southern University; NASA)*

Asteroids

While comets continue to pique everyone's interest, asteroids – their fellow Solar System wanderers – have earned a place in Hubble's pantheon of planetary science offerings. HST has imaged a number of these Solar System 'leftovers' in an effort to help planetary scientists understand them. One of the larger asteroids is Vesta, the most geologically diverse asteroid observed. It appears to have basaltic flows on its surface, indicating that it may once have had a molten interior (much as the Earth does). In December 1994 a team of scientists from the U.S.A. and European Southern Observatory used HST to image Vesta in four colors. The goal of the joint HST/ground-based observing campaign is to make a geochemical map of Vesta's surface.

The preliminary images showed Vesta to be a battered little place, having weathered a variety of collisions in the past that gouged material from its surface and sent it as meteorites to Earth. One huge collision created a 456-kilometer-wide impact crater. Vesta campaign principal investigator Ben Zellner of Georgia Southern University spearheaded the joint campaign. 'These observations showed that Vesta is far more interesting than most asteroids we see,' he said.

The reason for the team's interest is obvious when you see the asteroid: part of Vesta has been broken off, exposing the mantle of this tiny world. No other rocky body in the Solar System has an exposed mantle, and thus Vesta provides planetary scientists with a possible key to understanding the structure of larger planets in the Solar System. Despite this broken appearance, Zellner says that Vesta has remained essentially intact since the formation of the Solar System. 'Vesta provides us with a record of the long and complex evolution of the Solar System,' he said.

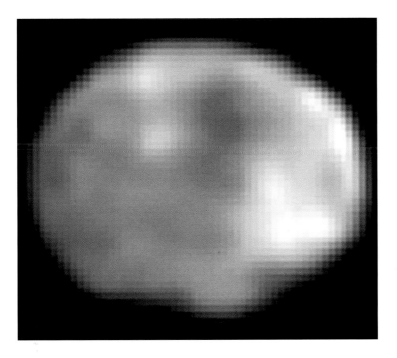

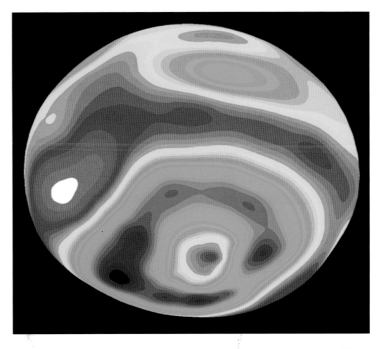

Elevation

-12km   +12km

Fig 2.39. (Upper left) An image taken in May 1996 when the Vesta was 110 million miles from Earth. The asymmetry of the asteroid and 'nub' at the south pole suggests that it suffered a large impact event. The image was digitally restored to yield an effective scale of 9.6 kilometers per pixel. **(Upper right)** A color-encoded elevation map of Vesta clearly shows the giant 456-kilometer-wide diameter impact basin and 'bull's-eye' central peak. The map was constructed from 78 WF/PC-2 pictures. **(Left)** A three-dimensional computer model of the asteroid Vesta synthesized from HST topographic data. The crater's 12.8-kilometer-high central peak can be seen near the pole. The surface texture on the model is artificial, and is not representative of the true brightness variations on the asteroid. Elevation features have not been exaggerated. *(Ben Zellner, Georgia Southern University; Peter Thomas, Cornell Univerity; NASA)*

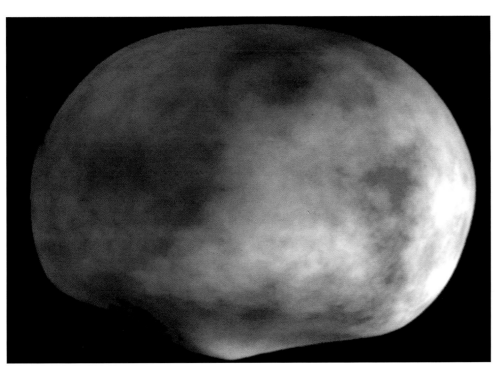

Other planets?

Ultimately, HST's studies of the Solar System bring us back to that long and complex process which formed the Sun and its planets. They seek answers to some of the most fundamental questions we can pose about our home in space: What caused it to form? What were the conditions in the early solar nebula? Everything we learn about our Solar System not only aids us in understanding our own place in the universe, but also helps us when we turn HST's gaze out to the stars – in search of planets around other stars.

Already HST has detected 'something' about 40 to 50 Jupiter masses around another star. As we shall discuss when we look at stellar evolution in Chapter 3, spectral analysis of the

Figure 2.40. This false-color image reveals the faintest object ever seen around a star beyond our Sun, and the first unambiguous detection of a brown dwarf. The brown dwarf, called GL229B, orbits the red dwarf star Gliese 229, located approximately 18 light years away in the constellation Lepus. The brown dwarf was first observed in far red light on October 27, 1994, using the adaptive optics device and a 60-inch reflecting telescope on Palomar Mountain in California. This image was taken with HST's WF/PC-2, in far red light, on November 17, 1995. *(T. Nakajima, Caltech; S. Durrance, Johns Hopkins University; NASA)*

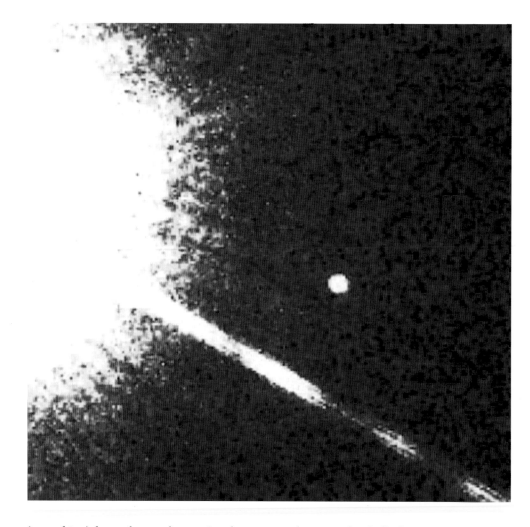

'something' shows that methane exists there – a gas that most decidedly does *not* exist in stars, but is in great abundance in the atmospheres of our own gas giant planets.

In the space around that star – and many others – we may be seeing the formation of solar systems similar to our own. HST's studies of our own Solar System have prepared us to see those worlds, but understanding them and their place in the universe requires us to turn our attention to the complex business of stellar evolution and the lives of the stars.

Stars and the interstellar medium

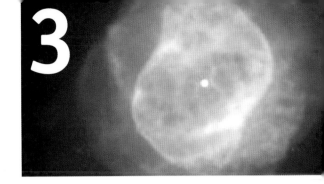

3

To be a star is to know eternal stress. To live as a star is to walk a never-ending tightrope, knowing that there can be only one outcome – your fall.

Martin Cohen

The next object I have observed is the essence or substance of the Milky Way. By the aid of a telescope anyone may behold this … the galaxy is nothing else but a mass of innumerable stars planted together in clusters.

Galileo Galilei

The lives of stars

Stars. They fascinate us. They are born, live their lives, and die – on timescales that make the longest human lifespan seem like only a moment in time. The physics of stellar evolution is a theme that echoes endlessly across the universe. The stars are variations on that theme, each one adding a distinct voice to the cosmic symphony. There are newborn stars and supernovae, hot supergiants and white dwarf stars, cataclysmic stars, and the ancient, slowly dying red dwarf stars. The understanding of all these variations forms an important recurring theme in the history of astronomy.

Stargazing prompts us to ask many questions. What are stars? How are they born? What happens when they die? Why are stars so different from each other? While we have begun to answer these questions in the past few decades, there is a lot of activity occurring in stars that we just do not understand yet. This may seem surprising, but remember that astronomers only recently developed the tools to interpret what stars are telling them. Looking at a sky full of stars is like wandering through a forest and seeing trees of all different shapes, sizes, ages, and states of health. There are tall trees next to short trees, and so one might decide that tall trees are older than short trees. That could be an erroneous conclusion since there are trees of different species in the forest, and some short trees might be older than their taller neighbors. Seedlings sprouting up here and there might lead one to conclude that they are weeds instead of baby trees. Dead logs and dried-out branches on the forest floor come from the surrounding trees, but an inexperienced observer might not be able to link specific branches to the surrounding trees, or determine why the trees and branches died. One might walk across piles of leaves and figure out that some leaves come from certain trees, but how to explain pine needles? Or pine cones? Or acorns? Or sap?

In short, to understand the rhythms of forest life, we have to know a lot about trees and their life cycles. It is the same way with stars, but, unfortunately, humans do not live as long as stars do. The best thing we can do is to make repeated observations of different kinds of stars and come up with theories to describe what we see.

So, what exactly *are* stars, anyway? The simple definition is that they are self-luminous spheres of gas. To truly characterize stars takes a detailed study of the different 'interpretations' of that definition. Of course, the best-understood example of a star is our Sun, which makes it a good place to start when studying the general properties of stars. Astronomers usually refer to other stars in terms of *solar* brightness, luminosity, mass, and radius. This gives us a nice shorthand way to refer to any stellar characteristics: L_S for luminosity; M_S for mass; and R_S to describe radius. So, for example, if a star's mass is given as $10M_S$, that means it has ten times the mass of the Sun.

Stellar masses range from about $0.1M_S$ to roughly $100M_S$. At the lower end, we are really testing the definition of the term 'star' because a star needs enough mass to produce temperatures that will sustain the nuclear reactions at its center. At the upper end, say stars that are $100M_S$, mass is harder to estimate, but there is no simple reason to assume an upper limit.

The radii of most stars range from $0.1R_S$ to $10R_S$. Exotic stars, such as white dwarf stars, neutron stars, and black holes also test the definition of the term star. Supergiants, such as Alpha Orionis (Betelgeuse) can be hundreds of solar radii across. White dwarfs are about the size of the Earth and may be one solar mass, while neutron stars are thought to be only about 30 kilometers across and can have a few solar masses. Stellar black holes, on the other hand, could be very small, with their exact size very dependent on how much mass has been swallowed up. A black hole of $5M_S$, for example, would have a size (that is, the radius of its event horizon) of about 5 kilometers.

Temperatures of stars are the only stellar measurement not given in solar units. Instead, they are given in kelvin (K), where 0 kelvin equals -273 degrees Celsius. Stars are classified by their temperature and luminosity. Luminosity is simply a measure of the amount of energy emitted by the star each second. The luminosity of the Sun, for example, is 4×10^{26} watts per second, and its temperature is 15 million kelvin at its core. Stellar surface temperatures range from 1000 kelvin for the coolest objects we call stars, to 5750 kelvin for the Sun, to super-hot stars measured at over 200 000 kelvin!

Astronomers plot the range of stars in a well-known chart called the *Hertzsprung–Russell diagram* (or H–R diagram for short). It is named after the two astronomers responsible for its introduction: Eijnar Hertzsprung and Henry Norris Russell, and its origin goes back many decades. It works simply by using stars as points on a plot, with the horizontal axis representing the effective temperature of a star and the vertical axis representing the star's luminosity. Temperatures can be estimated from the color of the star or by its spectral type – another form of classification that sorts stars by their chemical spectra.

Luminosity is estimated by using the star's absolute magnitude, the magnitude that the star

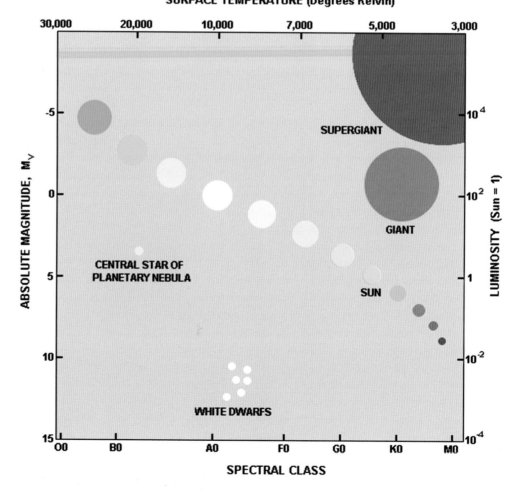

Figure 3.1. The Hertzsprung–Russell diagram is a graph of stellar brightnesses (labeled 'absolute magnitude') plotted against temperatures. The result is a classification of stars. Shown in unscaled sizes are the Sun and other stars of the main sequence (the band of stars that cuts across the graph from upper left to lower right), as well as a red giant, a supergiant, white dwarf stars, and the central star of a planetary nebula. *(T. Kuzniar, courtesy of Loch Ness Productions)*

would have at a standard distance of 10 parsecs. Since a star's apparent brightness changes with distance (specifically as the inverse square of its distance), this has the effect of putting all stars on an equal footing.

What fuels a star? The energy source for stars was established in the 1930s (primarily by physicist Hans Bethe) as nuclear fusion. The fuel for stars is primarily hydrogen, and when nuclei of hydrogen fuse to form the next heaviest element – helium – energy is given off in the form of heat and light. There are three major nuclear cycles that can operate in stars. At lower temperatures, the *proton–proton chain* transforms hydrogen into helium, and releases energy. The *carbon cycle* takes place in slightly higher-temperature stars. Hydrogen burning still occurs, but a carbon nucleus is needed as a catalyst. At the very highest temperatures, such as those found in red giant stars, the *triple-alpha process* transforms helium into carbon.

Study Figure 3.1 for a short while, and you can make some interesting observations about how the stars are grouped. The belt of stars slanting down and to the right is called the *main sequence*. Most stars spend their lifetimes as main-sequence stars, but there are some stars that never make it onto the main sequence. Above the main sequence lie the giant and supergiant stars. Below the main sequence is the domain of the dwarf stars. Some have evolved to their present position from the main sequence; others formed as dwarfs or giants. These designations also describe the radius of the star, and show that the measured radii increase as we go up and to the right on the H–R diagram.

A star's life on the main sequence is a relatively stable existence. After a somewhat stormy formation in a cloud of gas and dust (discussed in Chapter 2 and later in this chapter), the star settles down and spends most of its life happily converting hydrogen into helium. Unfortunately for the star, this cannot last forever. The process of nuclear fusion changes the composition of the core where it takes place, and this, in turn, changes the star. In the early stages of its life, the star is all hydrogen. Then, as nuclear fusion proceeds, a helium core grows. This stage of main-sequence life lasts until about 10 per cent of the star's mass has been converted into helium. The Sun, for example, has been on the main sequence for about 5 billion years, and should go on for another 5 billion years peacefully converting hydrogen to helium.

When the helium core becomes so big that the star is no longer stable, the fun begins. The core contracts, and this raises the temperature. At this point, helium burning begins to occur. This causes heavier and heavier elements, such as carbon and oxygen, to be created in the ongoing fusion process, and then these, too, are burned. In effect, ashes are being burned to squeeze out energy. A huge excess of energy in the core causes the star to swell into a 'giant' stage. At this point, the star leaves the main sequence. The fusion process continues until iron is created in the core. To burn iron would require more energy than would be created, and thus everything comes to a catastrophic standstill. In very massive stars, the core collapses and the outer part of the star blows away in an event called a *supernova explosion*.

When it comes to answering the questions of starbirth and stardeath, and all the variations in between, we have only to look at the sky. Everywhere we look, we can see the process of stellar evolution – places where stars are being born, where they are living, and the aftermaths of their deaths. Sometimes in their death throes stars initiate the births of other stars. In other places, colliding galaxies excite star formation – leaving behind a twinkling trail of stellar newborns.

The manner in which stars die intrigues astronomers. Often, each new discovery spurs more questions than it answers. This is not to say that scientists have not collected a lot of data, however. Sites of stardeath, for example, appear everywhere in the universe, and they remain fertile areas of study. How a star dies is largely dependent on the mass it ends with, rather than its original mass. Stars like the Sun spend most of their lives steadily converting hydrogen to helium. When the process ends, the star goes through the red giant phase. The end of nuclear burning signals the end of the giant phase. Then, the star shrinks, leaving behind a ghostly looking shell of gas surrounding it – a planetary nebula. Finally, the star cools and contracts, and becomes a white dwarf about the size of the Earth.

Stars slightly larger than the Sun – with final masses between $1.4M_S$ and $3M_S$ – also spend billions of years on the main sequence, converting their hydrogen to helium. When they contract, they can become *neutron stars*, so-called because the intense gravity at the core presses the protons and electrons together to form neutrons.

Stars with masses greater than $3M_S$ do not end up as neutron stars. They contract further to form a stellar black hole. The gravity is so strong that no light can escape the hole, although we do see radiation coming from hot material as it falls into the hole.

Stardeath in its many forms is one process that enriches the interstellar medium with chemical elements. Many stars lose mass throughout their lives, particularly in the form of stellar winds. Clearly, the most catastrophic mass loss they can experience comes when they die. As we will see later on in the chapter, supernovae eject essential elements such as iron into the universe, and these are found in the interstellar medium as concentrations of chemical elements. Eventually these elements turn up in other stars, planets, and, ultimately, our bodies. How can this be? Was everything not created when the universe came into being?

The Big Bang and the resulting Primordial Fireball produced the lightest elements: hydrogen, deuterium, helium, and a little lithium. For the production of heavier elements such as carbon, oxygen, calcium, and iron, however, ordinary stars are needed. Elements heavier than iron, such as lead and platinum, are produced in supernova explosions. The cycle of star formation out of interstellar gas and dust, and the subsequent re-depositing of stellar materials back into the interstellar medium during the various processes of stardeath, increase the amount of heavy elements in the universe over time. The material in the Sun and planets, for example, has been processed through at least one other star. We, having come to life here on Earth, also carry the atoms of long-dead stars in our bodies. Tracing the origins of these chemical elements back through the generations of stars that created them is one way of finding our own place in the universe.

This brings us back to HST, because its primary stellar mission is to gaze at a variety of stars, catching them in very distinct, and sometimes unusual, stages of existence. Before we look at HST's studies of starbirth, stardeath, and the denizens of its exotic stellar zoo, let us turn our attention to a unique part of space that we rarely think about as we gaze out to the stars – the interstellar medium, the 'empty' space where stars exist. It plays such an important part in the lives of stars that several HST researchers have devoted large amounts of time to probing its mysteries.

Chemical abundances in the interstellar medium

Imagine receiving an astronomy press release about stellar abundances in the mail. If you were a typical news editor, you probably would not read beyond the first sentence, unless the article happened to start off with the headline 'Hubble Gets The Lead Out – Telescope Finds Heavy Metals Between the Stars'. You might not expect to find such a humorous description of chemical abundances between the stars, but it turns out that there are enough heavy elements in the interstellar medium to warrant using valuable HST time to find them.

So, what is the excitement over finding heavy metal concentrations between stars? As we mentioned earlier, interstellar regions offer intriguing clues to the origins and amounts of chemical elements found in stars. If certain elements are especially 'abundant' in one area, then they should be plentiful in nearby stars. This, of course, assumes that the local stars formed in the same region and are not just 'passing through'. It is like walking through a forest and finding cabins built from the wood of surrounding trees. If there are more pine trees, the cabins will be built mostly of pine, while a scarcity of spruce trees means that fewer homes would have spruce wood.

Measuring chemical abundances between stars can also give astronomers a better idea of the life cycles of stars. As we have mentioned, older stars actually return much of their material back to the interstellar medium, and observations can detect that sort of activity. Newborn stars take in that material as they form. By probing the elements shed by stars, astronomers can follow the processes of star formation and death.

When scientists discuss the amounts of elements in the universe, they usually compare what they have observed to something called a *cosmic abundance*. Think of it as a sort of 'beginning inventory' of elements. Because not all stable elements are found in a single object, an estimate of the cosmic abundance is cobbled together from a variety of sources, including the Sun, other stars, nebulae, meteorites, and the Earth's crust. If astronomers find some amount of an element – say krypton – in a region of space, they can measure that amount against the overall abundances of krypton in the universe (it is only the *relative* abundance that is significant; the actual amount found is meaningless). Thus, the cosmic abundance of krypton becomes a sort of 'yardstick' against which astronomers compare specific measurements of krypton in various regions of space. If there is more krypton, then a source must be nearby.

To explore interstellar clouds of gas and dust in more detail, HST astronomers use the telescope's spectrographs, which are built to examine ultraviolet light. Prior to HST's deployment, the way to do good ultraviolet spectroscopy was to use spacecraft designed for it – the Orbiting Astronomical Observatories and the International Ultraviolet Explorer (IUE), for example. IUE did ground-breaking work during its 17 years on orbit, paving the way for HST's higher-resolution spectrographs to take much more detailed spectra of the interstellar medium.

One way to determine what is in the interstellar medium is to look at the light of stars shining through it. This is just what Ball Aerospace astronomer Dennis Ebbets and the late Jason Cardelli of the University of Wisconsin did when they used the Goddard High Resolution Spectrograph to determine the abundances of elements in the interstellar medium between Earth and a pair of stars. The stars they used as light sources to illuminate the intervening clouds of gas and dust were Xi Persei and Zeta Ophiuchus. Basically, the two scientists took advantage of the fact that certain elements absorb ultraviolet light. They examined the light dispersed with the GHRS and noted which lines indicated absorption. The heaviest element observed in the interstellar medium before the launch of HST was zinc. During HST's first three years on orbit, Ebbets and Cardelli took spectra that revealed the heavier elements lead, gallium, germanium, arsenic, krypton, and tin. 'We have now added eight elements that are heavier than zinc,' Cardelli reported. 'These elements are produced by specific nuclear processes in certain types of stars and dumped into the interstellar medium.'

Analyses of the spectra showed underabundances of several elements. Generally, when astronomers study interstellar abundances, they find that the amount of underabundance is greater for the heavier elements. The usual explanation is that the 'missing' elements *do* exist – but that they are 'hidden' as interstellar dust grains. This becomes clear by plotting the underabundances against each element's condensation temperature – the temperature at which the element becomes a solid. Since spectroscopy samples elements only in gaseous form, these 'missing' elements would not show up. In addition, the heavier elements form grains at higher temperatures, so they condense first in the high-temperature environment of star formation and therefore are much less abundant in the regions between the stars.

The exception to this is the noble gas krypton, which should, theoretically, show up all over the place. It does not form compounds with other elements very easily, and is unlikely to be incorporated into dust. In reality, astronomers only see about half the krypton they expect to find. Of course, this is puzzling. If it is not combining with anything else, where is it? The most reasonable explanation for this 'missing krypton' is that there is probably an error in the previously adopted cosmic abundance of krypton. It is now more likely that the amount measured is closer to the 'true' cosmic abundance. If this result is confirmed by other measurements of the interstellar medium with different stars, the cosmic abundance for krypton will have to be revised.

For Ebbets and Cardelli, this result was important, because it gives astronomers a more accurate feel for how much lead, krypton, and other heavy elements exist in the space between stars. 'The confirmation of these elements represents the first time they have been detected in the interstellar medium,' Cardelli said. 'It's like the promising start of a new day. We've opened a door to new explorations of the interstellar medium and the stars that dump material into it. With our insights we can explore how a galaxy processes and then mixes interstellar material.'

Fig: 3.2. Nebula in Orion as photographed by the 4-meter Mayall Telescope of the Kitt Peak National Observatory. *(National Optical Astronomy Observatories)*

Figure 3.3. (Opposite) A mosaic of the Orion Nebula, created primarily from WF/PC-2 images, augmented with groundbased data. The field of view stretches across 2.5 light years, encompassing a variety of star-formation activities including jets and proplyds. *(Bo Reipurth, European Southern Observatory; John Bally, University of Colorado; NASA)*

Newborns in space

The processes of starbirth, starlife, and stardeath take an incredibly long time. The best we can do, even with HST, is to take snapshots of as many stars as possible, hoping to piece together a coherent understanding of just how stars do what they do. Of all the processes astronomers study, the one that HST has really opened up to us is the stately progression of starbirth – from cold molecular cloud to planetary disks to complex plasma jets and finally to the fiery brilliance of a newly formed star.

The Orion Nebula

One of the Hubble Space Telescope's first views of the birth process of stars was in the gas and dust clouds in the Orion Nebula – a stellar nursery about 1500 light years away from Earth. The nebula lies below the three distinctive stars of Orion's Belt and looks, to the naked eye, like a faint greenish haze. Larger telescopes reveal an irregularly shaped cloud scattered with glittering stars. This area of the sky has been very rewarding for HST researchers, who turn the telescope's gaze toward it time and again. A brief look at the physics of gas clouds and hot young stars provides us with an understanding of what is happening in this region.

When we look at the Orion Nebula, we are seeing a diffuse nebula called an HII region. (HII is ionized hydrogen.) Most HII regions are heated by nearby stars, and they produce bright emission-line spectra. Of course, *we* see visible light when we gaze at the Orion Nebula through

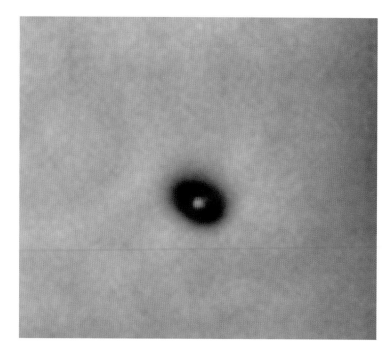
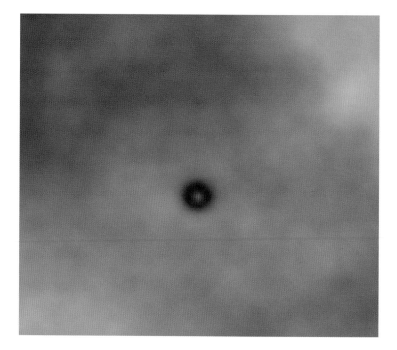
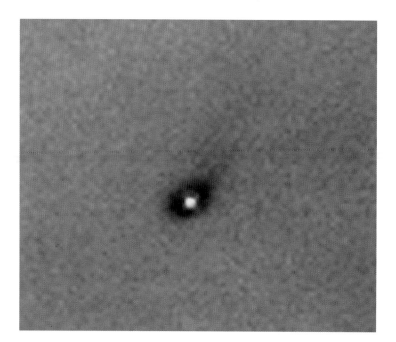
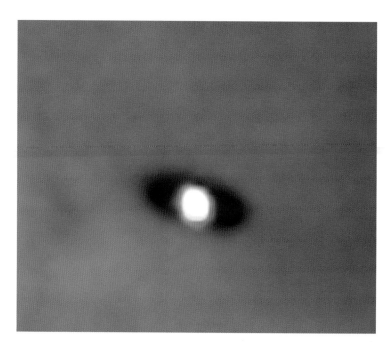

Figure 3.4. Protoplanetary disks, or 'proplyds', in the Orion Nebula, seen approximately face-on. The orientation of each system accounts for their appearance in this image. The red glow in the center of each system is a newborn star (approximately one million years old) shining through the disks of gas and dust. The dark disks are silhouetted against hot gas in the nebula. Each proplyd is about 30 times the diameter of our Solar System. Neutral hydrogen is indicated by green, red light denotes once-ionized nitrogen and the blue light indicates twice-ionized oxygen. *(M. McCaughrean, Max-Planck Institute for Astronomy; C.R. O'Dell, Rice University; NASA)*

binoculars and small telescopes, whereas HST sees radiation over a wider range of wavelengths. Each gas in the nebula radiates at slightly different wavelengths, so if you know which region of the spectrum is specific to each gas, you can construct a 'color map' of the nebula based on the wavelengths you see. In Figure 3.3, light from neutral hydrogen is coded green, red light is the code for once-ionized nitrogen, and the blue light indicates twice-ionized oxygen.

The Orion Nebula is a good example of what astronomers call 'blister' nebulae. It is an illuminated region on the edge of a dark cloud of material called the Orion Molecular Cloud (OMC). The blister is the area where the ultraviolet light from young stars is lighting up the surface of the cloud. The stars in the cloud are around 300 000 years old – which makes them mere babies in the cosmic scheme of things.

The search for hot, young stars in the Orion Nebula has led astronomers such as Robert O'Dell to study the region in great detail. More than 150 young stars have been surveyed here. Several have formed great plumes of gas, blowing matter away from themselves and tearing into

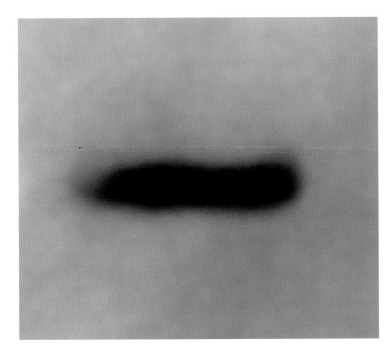 

the surrounding nebula at 160 000 kilometers per hour. The resulting shock waves 'eat away' large scalloped regions of the nebula.

How can something as diffuse and tenuous as an interstellar cloud of gas and dust give rise to hot young stars? Starbirth is inevitable in interstellar hydrogen clouds. It may take millions of years before a star is born, but it will happen. Molecules of gas and dust grains move around and bump into each other in a process called 'mixing'. Give these cold molecular clouds enough time, and maybe a little push from the outside, and they will start to form a star. That little push may be in the form of shock waves from a nearby supernova or the action of a passing star as it plows through the cloud. A region of higher density starts to form a disk-like shape and material falls toward the center. The cloud spins faster, just as when an ice skater spins faster and faster during a pirouette. At some point, the temperature in the center becomes sufficiently high enough for a star to 'turn on'.

Once the star is born, a great deal of material in the disk is still present, spinning around the star and reflecting starlight. In some cases, material falls onto the star and is ejected out in a plume. If there remains enough gas and dust, eventually *protoplanets* will form in regions of higher density within this *protoplanetary disk*. The protoplanets will continue to accrete material, and over time, create something approximating our Solar System. Anything remaining from that process would inhabit the farthest reaches of the new system, as the Kuiper Belt and Oort Cloud of comets do for our own Solar System.

Now, with all this starbirth going on in the universe, one might think that astronomers would see disks around every newborn star. They do not, but this does not mean that the disks are not there. These accretion disks are very hard to see because at some stages they are composed mainly of cold gas and dust particles, which are easily outshone by the nearby star. These disks are usually more detectable in the infrared than in visual or ultraviolet wavelengths. O'Dell considered himself lucky to find them in the Orion Nebula for a couple of reasons – they are near the edge of the nebula, and they are being lit by hot young stars that have only recently 'turned on'. Because they are so hot, these stars are great emitters of a lot of ultraviolet light, which lights up the surrounding cloud of gas, and, in turn, the gas emits light.

If the stars are lighting up gas clouds in their region of the Orion Nebula, it stands to reason that they will also illuminate clouds of denser material that are the birthing grounds for planets. Figure 3.4 shows a face-on view of these protoplanetary disks – or 'proplyds' as they have been dubbed – in the Orion Nebula, and Figure 3.5 shows an expanded, edge-on look at one. Proplyds were a serendipitous discovery. According to O'Dell, he was really studying the

Figure 3.5. A Proplyd seen edge-on in the Orion Nebula. This disk is approximately 17 times the diameter of our Solar System. The left-hand image was taken with the same color coding as figure 3.4, and the right hand figure was taken through a filter that blocks the emission lines from the nebula. The bright nebulosity above the disk betrays the existence of the central star, although the star is not visible. *(C.R. O'Dell, Rice University; NASA)*

Figure 3.6. A generic model for the proplyds. A core of about 50 astronomical units across is ionized and heated from the outside. Gas and dust (arrows) flow away from the hot, outer layers of the core, and these interact with the surrounding nebular material. The exact appearance depends on the distance to the ionizing/heating star and the viewing angle. *(C.R. O'Dell, Rice University; G. Dinderman, Sky Publishing Corporation)*

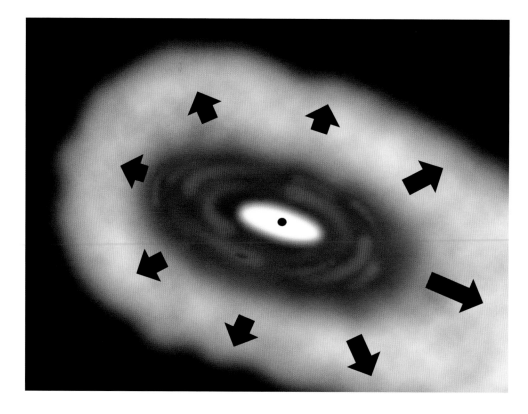

dynamics of star formation in the Nebula. In the process, he also found large flattened disks of material clumped around them. 'Those disks are a good example of something we should have expected to see,' he said. 'Once we understood what they were, it's obvious that we should have been looking for them all along.'

The generic model for proplyds in Figure 3.6 illustrates the mechanics of the situation. There is a core region about 50 astronomical units wide, illuminated by a hot star (represented by a black dot in the center). The dark ring-like region around the star is the part of the system swept clear of the gases that make up the star – and the region where rocky planets are most likely to form. The outer ring of material is being blown away from the star, and formation of gas giant planets would take place there.

Since starting their exploration of the Orion Nebula, O'Dell's team has found more than 150 proplyds in the region. No one expects to see actual planets embedded in these pancake-shaped disks of dust and gas – they would be far too small and faint to be detectable in images. However, many of these proplyds appear to have enough material to form planets someday. The masses of the disks range from one-tenth that of the Earth's to more than 700 Earth masses!

The Eagle Nebula

The intricacies of star formation occur all over the sky. One of the most famous examples lies 7000 light years away in the constellation Serpens. The Eagle Nebula, also described by Burnham as 'The Star-Queen Nebula' because it reminded him of a celestial throne, is a breathtaking sight through a backyard telescope. Under the gaze of the Hubble Space Telescope's WF/PC-2, it takes on epic proportions.

The images sent by HST show large columns of cold gas and dust stretching across the sky. The interstellar gas in these columns is very dense, exactly what we might expect of a stellar birthplace. The light from nearby young stars illuminates the tops of the towers, and exposes them to ultraviolet light. The ultraviolet radiation heats and erodes the tower tops. If there are stars forming here, they are still hidden inside the clouds.

Jeff Hester of Arizona State University, Tempe, Arizona, did the original research and analysis of the Eagle Nebula (also referred to as M16). He interprets the process the clouds are

undergoing as being a bit like a wind storm in a desert. 'As the wind blows away the lighter sand, heavier rocks buried in the sand are uncovered,' he said. 'In M16, instead of rocks, the ultraviolet light is uncovering the denser egg-like globules of gas that surround stars that were forming inside the gigantic gas columns.'

This process of evaporation eventually uncovers the young stars for our view, and illustrates processes that can limit a star's growth. For example, a young star surrounded by gas and dust, and isolated in its birthplace from other stars, grows until its mass triggers nuclear reactions in its interior. Then, the star begins to radiate light, and eventually mass would begin to flow away from the star in the form of a narrow jet and eventually a stellar wind. The 'mass loss' from a newborn star cleans out the local environment, which effectively stops the star from

Figure 3.7. The Eagle Nebula, M16. Giant pillars of gas and dust are lit by a bright star (just above the image area). The intense light and radiation given off by the star cause the gas to evaporate, and the process cuts off the newly forming stars from the gas clouds that feed them. *(J. Hester and P. Scowen, Arizona State University; NASA)*

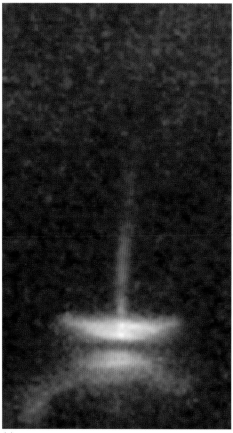

(a)

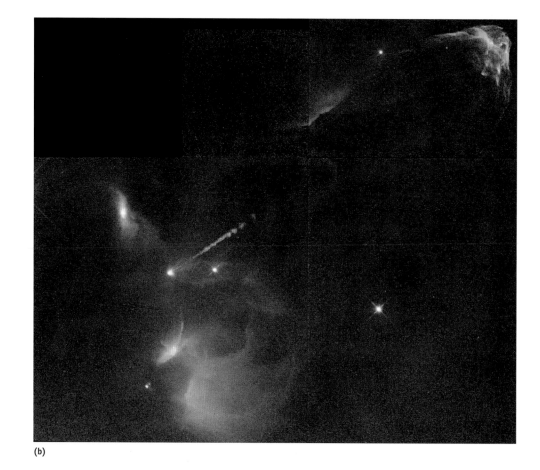

(b)

Figure 3.8. Jets from newly forming stars. The image of HH30 **(a)** shows the dusty disk edge-on, obscuring the star (in the dark lane between the two bright parts of the disk). The jets stretch out perpendicular to the plane of the disk. HH34 **(b)** shows a jet with a beaded appearance, which suggests that material may be ejected from the region in pulses. The jet in HH47 **(c)**, reveals a complex pattern that indicates that its star (hidden inside a dust cloud near the left edge of the image) could be wobbling (possibly from the pull of a companion star). Finally, the scales in the images (see Figure 3.9) show the large distances involved in this most explosive part of star formation. *(C. Burrows, STScI; J. Hester, Arizona State University; J. Morse, University of Colorado, STScI; NASA)*

(c)

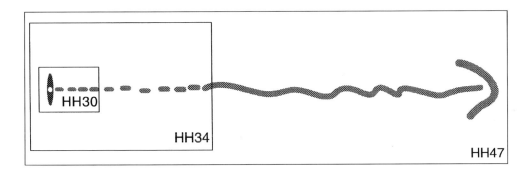

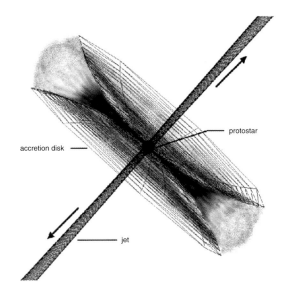

Figure 3.9. Distance scales involved in the stellar jets emitted from HH30, HH34, and HH47. The HH30 field shown is about 240 billion kilometers across. The HH34 field shown is 720 billion kilometers across, and the HH47 field shown is over one trillion kilometers across. (This is roughly 170 times the distance from the Sun to outermost Pluto in our own Solar System.) The arrowhead-shaped bowshocks created when a jet plows through interstellar material are also seen for HH47. *(STScI)*

Figure 3.10. Schematic diagram of the HH30 circumstellar disk. *(STScI)*

getting any bigger. However, the evaporation here has already cleared out the area and the star's mass is limited by this.

Also, because the evaporative process eats away all the material, it is not very likely that planetary systems will form around these stars. Indeed, the disks where planets form will never even have a chance to form. In the Eagle Nebula, only stars will be left after the gas is eaten away, and, as Hester notes, 'If these disks haven't formed yet, they never will.'

An alternative interpretation is that this region is a place where the process of starbirth failed, leaving only the dense clouds of material. Much of the gas cloud originally available for star growth has been eaten away by evaporation. The stars that do manage to form and shine here in the future will have less material to use as they form, and will be less massive.

Herbig–Haro objects

Long before HST was built, astronomers had noticed knots of nebulosity in the Orion Nebula. George Herbig of the Lick Observatory in California, along with Mexican astronomer Guillermo Haro, cataloged these objects independently in the early 1950s. At first regarded as curiosities, it turns out that these so-called Herbig–Haro (HH) objects play an important role in star formation. Insight into their contribution to the very early stages of star and planetary system formation has been provided by a powerful combination of detailed, incisive views from WF/PC-2 and wide-field images obtained with sensitive ground-based CCDs. These new results confirm the physics of the formation of dusty disks around newly forming stars and show that violent processes with effects seen over immense distances are an integral part of the births of stars.

Hubble's views of three Herbig–Haro objects, seen in Figures 3.8 and 3.12, illustrate the dramatic discoveries. The first, HH30, lies some 450 light years away in the constellation Taurus. This protostellar object is viewed nearly edge-on in Figure 3.8 and shows a dust disk around the newly forming star. The central region of the disk itself is the darker slot between the two white arcs. The star is hidden within the disk, but its light illuminates the top and bottom parts of the disk. The reddish column shows the jet that originates close to the star.

Here, in the case of HH30, we see this stage of stellar evolution essentially laid out for our inspection, and it is quite clear that the object provides the 'smoking gun' evidence needed to confirm the nebular model of star formation. It is so definitive that astronomer Chris Burrows of the Space Telescope Science Institute felt compelled to describe it as a major breakthrough. 'For the first time,' he said, we are seeing a newborn star close up – at the scale of our Solar System – and probing the inner works.'

Recall that during star formation, when the rotating disk of gas collapses by self-gravity, the central blob eventually becomes the new star. The leftovers in the disk serve as the raw material for the formation of planetary system bodies. The existence and extent of jets emanating from newly forming stars is a development that – as one scientist has put it – astronomers should have expected, but didn't. Why jets from newborn stars? In the accreting nebula, material from the disk falls onto the star and creates hot material that blasts back into the interstellar medium via these jets, as illustrated in the schematic (Figure 3.10). The jets retain a narrow,

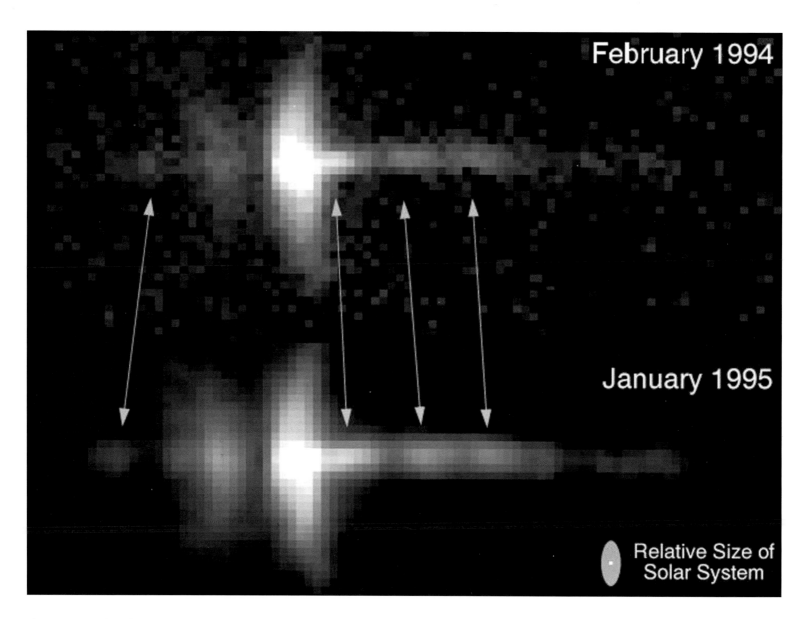

Figure 3.11. Motion of jet material in HH30. Comparison of WF/PC-2 images taken in February 1994 and January 1995 shows the movement of clumps of material in the HH30 jet. *(C. Burrows, STScI; NASA)*

Figure 3.12. (Opposite) Stellar jets in HH1 and HH2. The top image, taken with the WF/PC-2, shows the entire jet stretching across 1 light year of space. The close-up view at lower right shows clumps of ejected gas, and (lower left), the by-now classic arrowhead shape of a jet/bowshock collision. *(J. Hester, Arizona State University; NASA)*

collimated structure as they blast many billions of kilometers across space from the star. Clearly, some mechanism prevents them from expanding or opening up. Astronomers suspect the interplay between stellar magnetic fields and the inherent fields in the nebula may be corralling the jets into the narrow fields we see.

In HH34 we can see the distances that these jets reach out from their stars. The jet extends out more than 700 trillion kilometers, and has a clumpy look about it, almost like beads on a string. Material is being ejected episodically, perhaps triggered when chunks of material from the surrounding disk fall onto the star, lending the entire jet its beaded appearance. The speed of these jets is enormous – approximately 800 000 kilometers per hour! As we shall see from images of other HH objects, this velocity is sufficient to carry the jet material across vast distances of space from the star.

The expansion of scales continues with the image of HH47, where the jets are plowing into clouds of interstellar material. The general view is that of a jet emanating from a circumstellar disk (HH30), propagation as a narrow jet over large distances (HH34), and the eventual collision with interstellar material producing arrow head-shaped structures (HH47). The scale of this image is more than one trillion kilometers across, illustrating the powerful reach of these streams. The young star (hidden by the dust cloud near the left edge of the image) is producing this long jet, and partially reveals itself by the white filaments that shine in the star's reflected light. The end of the jet at the right shows the collision region where the outflow

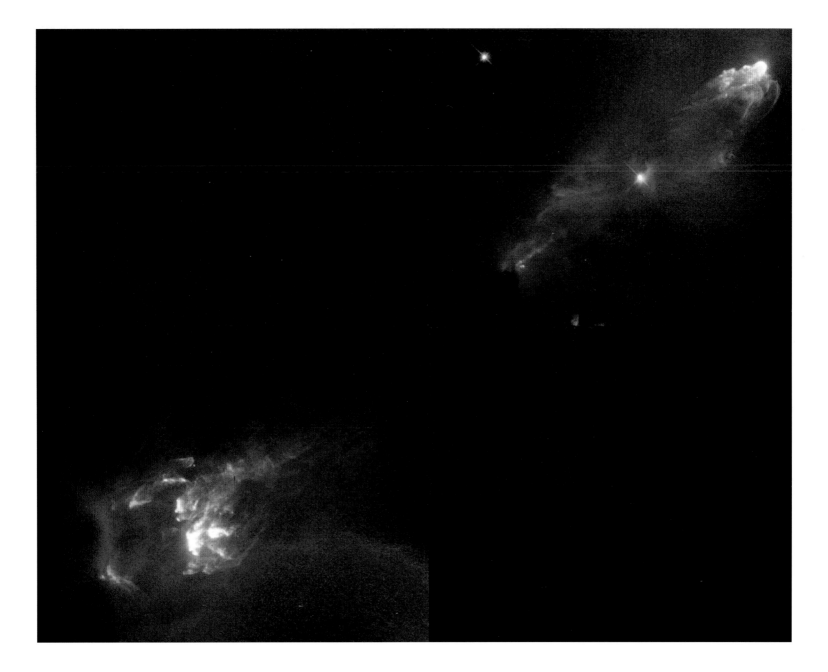

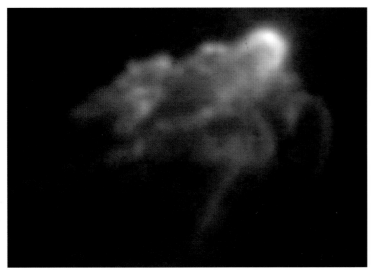

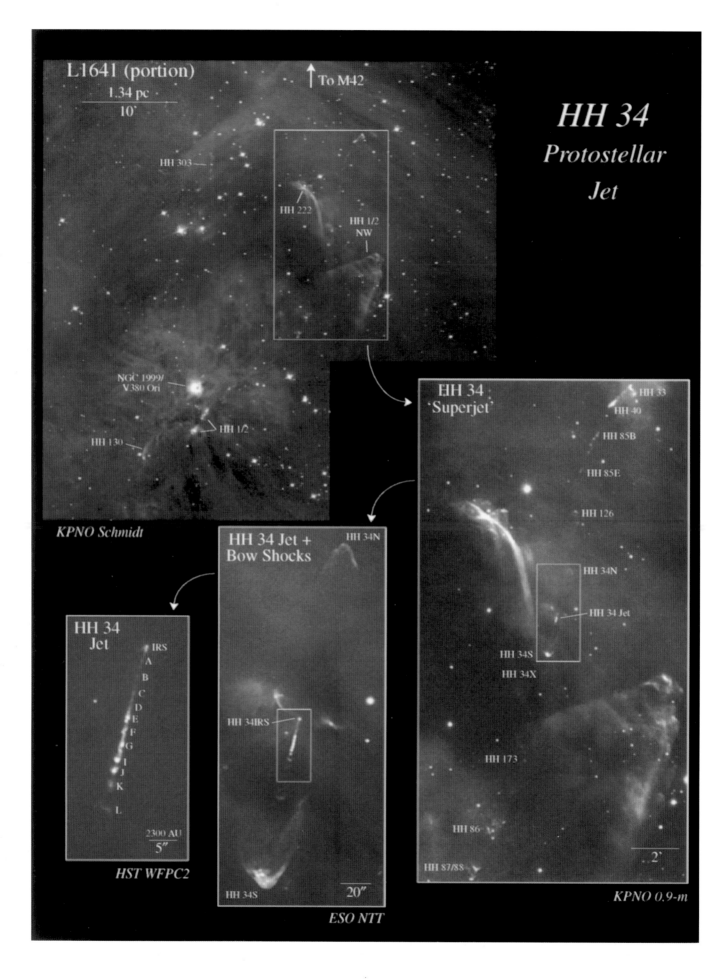

L1641 (portion)

1.34 pc / 10'

To M42

HH 303

HH 222

HH 1/2 NW

NGC 1999/ V380 Ori

HH 130

HH 1/2

KPNO Schmidt

HH 34
Protostellar
Jet

HH 34 'Superjet'

HH 33

HH 40

HH 85B

HH 85E

HH 126

HH 34N

HH 34 Jet

HH 34S

HH 34X

HH 173

HH 86

HH 87/88

.2'

KPNO 0.9-m

HH 34 Jet +
Bow Shocks

HH 34N

HH 34IRS

HH 34S

20"

ESO NTT

HH 34
Jet

IRS
A
B
C
D
E
F
G
I
J
K
L

2300 AU / 5"

HST WFPC2

from the star slams into the gas cloud. The ionized gases glow in an eerie light, which also reveals the complex interactions between the outflow and the cloud.

This view is by no means unique to the HH objects just discussed. Figure 3.12 shows images of HH1 / HH2 in the constellation Orion. The entire jet (left) is slightly more than a light year across from tip to tip, and is a good example of a jet where both ends are visible. The star itself lies midway between the ends of the jet, and the region around the star shows the jet with the beaded appearance (close-up below). The detail of the classic bowshock (arrowhead structure) produced when the jet material plows into interstellar clouds is shown at upper right.

A more complete picture of these HH regions comes from the marriage of ground-based and HST images. When these studies are combined, we can get a good idea of the true complexity of the process of starbirth. Astronomers John Bally and Jon Morse, of the University of Colorado, teamed up with astronomer Bo Reipurth, who was at European Southern Observatory, to combine data from HST with wide-field images from sensitive CCDs in South America. The marriage is a good one because the complex of structures associated with many HH objects can extend across many parsecs of space. The HST can zero in on small fields of view, while the ground-based instruments can give us a very wide field of view of the same regions.

Figure 3.13. (Opposite) Details of the HH34 protostellar jet, in ground-based and HST images. Emissions are from hydrogen and ionized sulphur. The instruments are: **(upper left)** Kitt Peak National Observatory (KPNO) 0.6-meter Schmidt; **(lower right)**, KPNO 0.9-meter telescope; **(lower middle)** European Southern Observatory 3.5-meter New Technology Telescope; **(lower left)** HST WF/PC-2. *(J. Bally, J. Morse, and B. Reipurth, University of Colorado; NASA)*

Figure 3.14. Details of the HH111 protostellar jet. The top image is a 1 degree field-of-view image from the Cerro Tololo Inter-American Observatory 0.6-meter Curtis Schmidt. The bottom image is from the WF/PC-2. *(J. Bally, J. Morse, and B. Reipurth, University of Colorado; NASA)*

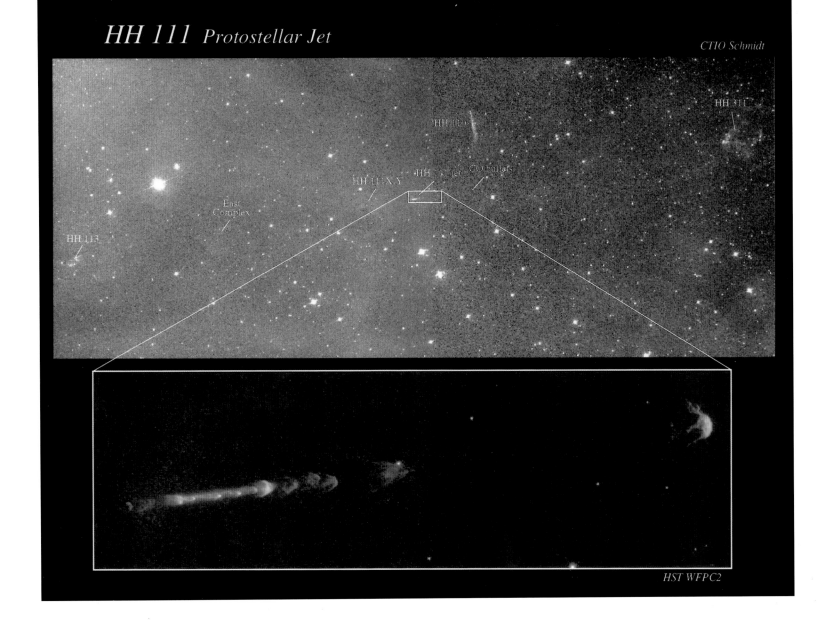

This is well illustrated in the set of images that make up Figures 3.13 and 3.14. The left-hand image shows the overall view of the HH34 system. Included are more detailed views, including the HST image. Concentrating on the overall view, we see that HH34 is only the central part of a chain of bowshocks. Moving to the south (toward the bottom), we see HH34S, HH34X, HH173, HH86, HH87, and HH 88. Moving to the north (upward), we see HH34N, HH126, HH85E, HH85B, HH40, and HH3. The system is about 2.7 parsecs long and S-shaped. It is clear that the direction of the jet has wobbled by 5 to 10 degrees over the last 5000 years, producing the delicate-looking S-curve. The bowshocks within the flow are all moving away from the jet.

Another example of the complexity and large distances is the HH111 system shown in Figure 3.14, which combines an HST image with a Curtis Schmidt image from the Cerro Tololo Inter-American Observatory. The central star is obscured, but the jet is visible in the HST image. Examination of the Schmidt image shows an outflow that ends at HH311, westward (right) and HH113 eastward (left). The shocks are moving away from HH111 and the entire system is about 7 parsecs long. Radio wave data show a chain of carbon monoxide (CO) 'bullets', not visible in the optical wavelength images.

These recent developments have provided exceptional new insights into the star formation process. The physical picture of accretion disks with jets of material – once considered a theoretical model – has now been seen in intricate detail. Continued observations and new discoveries in these regions will refine the current model even more. Certainly, the outflow of material in jets appears to be an integral part of the star formation process, and a major surprise is the distance to which these jets extend, sometimes across many parsecs. When we recall that the nearest star to the Sun is about one parsec away, the extent that these jets can influence the star-forming regions of our galaxy is impressive indeed. As astronomer John Bally put it, HST is bringing much of this activity into sharp focus, and giving us a greater understanding of the youth of the cosmic forest we seek to study. 'We tend to concentrate on leaves of the trees,' he said. 'Now, with HST, we have a scalpel that allows us to diagnose the details of the jet structure, the shocks in the flows. Now, we must learn what the detailed structure of the jets is telling us about the action at their source – at the newborn star.'

Beta Pictoris

Disk formation and production of planetary-system-like bodies are features of most theories of planetary system formation. Now that we have seen evidence for them, we know that protoplanetary disks around stars are no longer theoretical models. In fact, astronomers have seen many in ground-based images and radio telescope surveys, as well as from instruments with

Figure 3.15. This is the best ground-based image of the disk around Beta Pictoris. The disk around the star appears four times thicker in this ground-based image, due to the limitation of atmospheric seeing. This was obtained at the University of Hawaii's 2.2-meter telescope on Mauna Kea. *(Paul Kalas, University of Hawaii)*

infrared capabilities. The best-known protoplanetary disk was found around the star Beta Pictoris using a telescope at Hawaii's Mauna Kea observatory, and it was subsequently studied by the Infrared Astronomy Satellite. Beta Pictoris looks like a star with a cloudy ring of material around it. HST zeroed in on the star – which lies just 54 light years away in the constellation Pictor. We see HST's various views of Beta Pictoris in Figures 3.16 and 3.17.

The Goddard High Resolution Spectrograph was programmed to complete a spectroscopic study of Beta Pictoris. The resulting spectra tell us that there *are* clumps of gas falling into the central region. In fact, both the GHRS data and earlier IUE observations suggest that 100 to 150 clumps of material fall into the star each year. Figure 3.18 is a schematic attempt to describe this newborn system. With the Orion proplyds and the Beta Pictoris disk, astronomers may be seeing different stages of the same process of star system formation.

Figure 3.16. The thin disk of dust around the star Beta Pictoris. This WF/PC-2 image indicates that the thickness of the disk is no more than 600 million kilometers. A thin disk is a likely location for the formation of comet-sized or larger objects. *(A. Schultz, Computer Sciences Corporation; NASA)*

Figure 3.17. A tilt in the disk around Beta Pictoris, caused possibly by a large unseen planet. Compare the position of the white and pink parts of the image to the dashed line which matches the position of the outer disk – the red, yellow, and green parts of the image. The white and pink parts are below the line at left, and above the line at right. Thus, the inner part of the disk is tilted by the presence of an unseen object. *(C. Burrows and J. Krist, STScI, WF/PC-2 Team; NASA)*

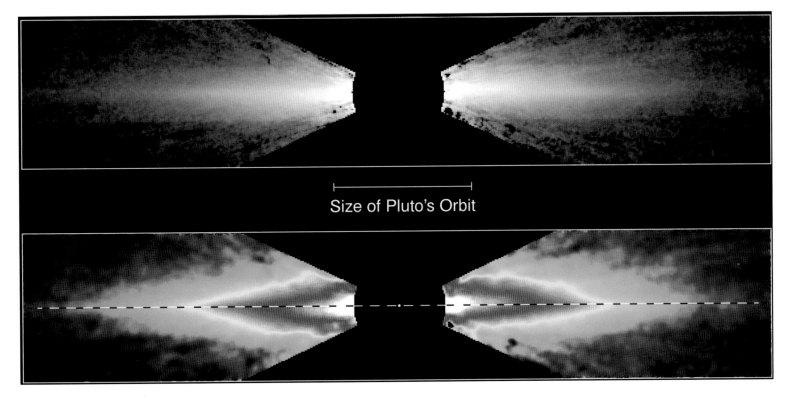

Size of Pluto's Orbit

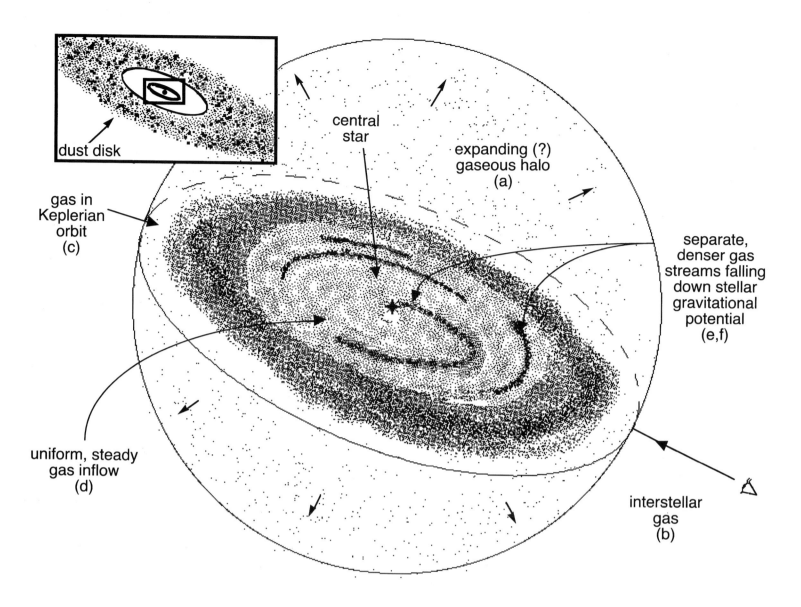

dust disk

central
star

expanding (?)
gaseous halo
(a)

gas in
Keplerian
orbit
(c)

separate,
denser gas
streams falling
down stellar
gravitational
potential
(e,f)

uniform, steady
gas inflow
(d)

interstellar
gas
(b)

Figure 3.18. Possible structure is: (a) an expanding halo of gas; (b) the interstellar gas around Beta Pictoris; (c) the gas in (Keplerian) orbit around the star; (d) uniform, steady gas inflow; (e) and (f) denser, separate gas streams falling down (the stellar gravitational potential) toward the star. *(Dara Norman, Computer Sciences Corporation, University of Washington, and the GHRS Science Team; Allen Home, Catholic University of America and the GHRS Science Team)*

A unique explanation of the nature of the infalling clumps of material has been suggested by French astronomer Anne-Marie Lagrange, of the Laboratoire de Astrophysique, Grenoble, France. Her interpretation is that these clumps begin as large comets on orbits that take them close to the central star of the system. When they sublimate, the gas that we see in the spectra is created.

Additional images of the Beta Pictoris disk obtained from WF/PC-2 clarify and hint at the presence of planets that have already formed within the cloud. In Figure 3.17, an HST image and a comparison ground-based image, the disk, which is composed of microscopic particles of dust and ice, is seen nearly edge-on. Both images show a thin disk, but the HST image has better spatial resolution. It shows a much thinner disk – about 25 per cent less thick than groundbased data have shown. A thinner disk means a higher density of material and increases the likelihood that comets and planets have formed by accretion in the disk.

If the current interpretation is correct, these images support the existence of a roughly Jupiter-sized planet in orbit around Beta Pictoris. Basically, the disk is not completely flat as might have been expected. The investigation team was led by Space Telescope Institute scientist Christopher Burrows, who found surprising evidence that the innermost region of the disk is orbiting in a different plane from the rest of the disk. (See Figure 3.17.)

The inner, pink coded region is clearly tilted (down at left, up at right). Such a tilt cannot last very long, so the perturbing body must still be twisting or warping the disk. It has been

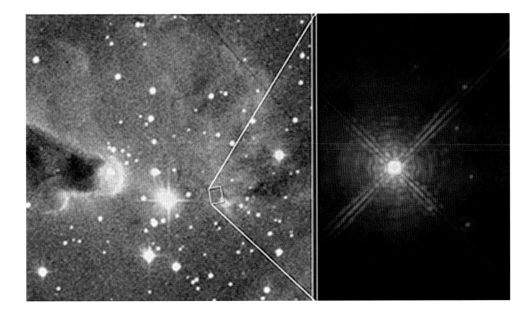

Figure 3.19. Starbirth in the Cone Nebula. The optical ground-based image on the left shows the 'cone' feature, with the region where the NICMOS zeroed in on a massive star and its six newborn companions (right). *(Rodger Thompson, Marcia Rieke, and Glenn Schneider, University of Arizona; NASA)*

estimated that a giant planet in orbit around the star would affect the disk as we see it. The planet's orbital tilt would need to be some 3 degrees off from the plane of the Beta Pictoris disk. Other spectroscopic studies of the star seem to confirm this planetary hypothesis, although a larger-scale warping of the disk seen in recent HST images could also be caused by the gravitational influence of a passing star.

Beta Pictoris is the only known mature star (that is, a star on the main sequence) that can be optically imaged and studied as described. While the research here gives us exceptional insight into star and planet formation, more stars like Beta Pictoris need to be studied in order to confirm that it represents a typical phase in stellar evolution and planetary system formation. The Near-Infrared Camera and Multi-Object Spectrometer (NICMOS) provides the near-infrared capability for such a search program.

Cone Nebula/NGC 2264

Sometimes special circumstances create conditions favorable for star formation, and the infrared capabilities of NICMOS have revealed an example in the region of the Cone Nebula. Figure 3.19 shows a visible light image of the region (left) and an infrared image of a small part of the visible light image, marked by the box (right). This star nursery lies about 2500 light years away in the constellation Monoceros. The brightest star here is NGC 2264 IRS, which is not seen in the visible image, but is very bright in the special 1.1-, 1.6-, and 2.2-micron wide filters that NICMOS uses to study light.

Mass loss in the form of stellar outflows from this star produced a cone of cold molecular hydrogen and dust, as seen on the left side of the visible image. These same outflows produced dense regions of gas and dust nearby. In infrared, this very bright, massive star stands out, along with six newborn stars. These stellar infants are only 0.04 to 0.08 light years away from their brilliant mother star. NICMOS's ability to peer into the dust around these stars has allowed us to see this stellar family of newborns whose formation was triggered by the massive star.

HST and the stellar zoo

When it comes right down to it, the stars HST looks at fall into three categories: those that are *getting ready* to do something, those that are doing something *right now*, and those that have *just finished* doing something. The telescope is rarely programmed to look at stars which make up a fourth category: those that live their lives quietly, doing nothing spectacular for millions of years. The reason is obvious – it is the same reason that spectacular events grab headlines in the media. No one is really interested in seeing stories that proclaim, 'Entire City Goes About Its Business'. However, if one morning the newspaper headline read 'Entire City Disappears Overnight', it *would* catch our attention.

With stars, the headline-grabbing stories cover such exotic things as the starbirth regions we have just discussed, the evolution of supermassive stars, the deaths of violently exploding stars, the complex interaction between double stars, and, of course, stars that are not like anything we have seen before. Those stories describe stellar activities that would command our attention if we did not live in a blissfully happy state of existence around a medium-sized yellow star in a fairly quiet stellar neighborhood. We are safe in our relatively uneventful part of the galaxy, but all around us are neighborhoods of stars whose lives make the Sun look quite bland indeed.

Supergiants, stellar winds and recycled stars

The Large and Small Magellanic Clouds are a familiar sight to southern hemisphere stargazers. They are actually companion galaxies to the Milky Way, and lie about 170 000 light years away from us. The Large Magellanic Cloud is home to a collection of hot, massive young stars called blue supergiants. Astronomer Sarah Heap has long been fascinated with the characteristics of these stars. Of particular interest is why these stars form in such massive proportions. Heap and her colleagues analyze the ultraviolet light streaming from those stars, using spectra taken with the GHRS.

Melnick 42

Figure 3.20 is the spectrum of one such star called Melnick 42, a roughly $100M_S$ star about 2.5 million times brighter than the Sun. It takes a little work to interpret the spectrum of Melnick 42, but it is well worth the effort. Notice the shape of the spectral line marked 'ionized carbon', which comes from carbon that has lost three electrons. This is the typical spectral signature of a star that is losing mass, so we know that Melnick 42 is blowing away large amounts of its gaseous outer envelope. This characteristic profile is also called a *P Cygni profile* – after a star in the constellation Cygnus that has this readily recognizable type of stellar wind. While all stars blow material away from themselves, supergiants blow away proportionally more than smaller stars, and at a higher velocity. How fast does this wind blow out from the star? To find out, Heap and her colleagues compared the profile of Melnick 42 to a detailed computer model for a stellar wind. It appears that the wind is blowing away at about 2900 kilometers per

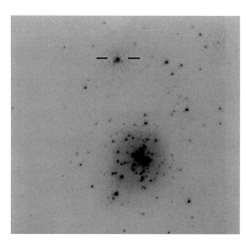

Figure 3.20. WF/PC-1 and GHRS observations of Melnick 42. The characteristic shape of the line marked 'ionized carbon' is shown at right. Emission (brighter) toward the longer wavelengths and absorption (darker) toward the shorter wavelengths – comes from gas flowing away from the star in all directions being illuminated by light from the star (shown bracketed above). See the text for a discussion of the analysis. *(S. Heap, NASA-Goddard Space Flight Center)*

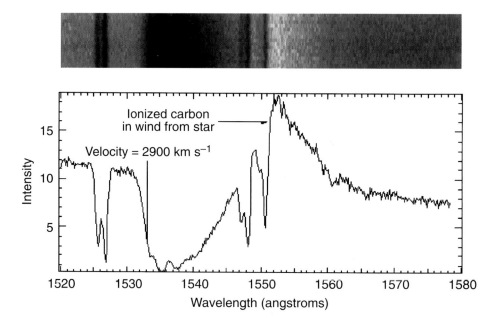

second. This means that the star is losing mass at a rate of $4 \times 10^{-6} M_s$ per year. In other words, if the star continues with this rate of loss, Melnick 42 would lose the equivalent of four Suns every million years. By comparison, this is about 100 million times stronger than the mass loss in the wind streaming out from our Sun.

Using other detailed models and HST observations, Heap came up with an independent check on the mass of Melnick 42. The WF/PC-1 image of the area shows no evidence for more than one star (it would confuse things if there were several stars there and you did not know it). Once Heap catalogued the characteristics of this hot young supergiant star, she turned her attention to others in the same region and took spectra of another star in 30 Doradus, the cluster where Melnick 42 is found. Since then, she has taken pictures with the Wide Field and Planetary Camera and has found that the whole region is a cluster of stars. If you could take all the stars in the region and lump them together, you would have a conglomeration of 100000 solar masses.

What Heap thinks she has really found here is another stellar nursery of hot young super-giants. 'This is fantastic to find in another galaxy,' she said. 'It's kind of your "prototype starburst" – and it is close enough so that you can actually resolve stars. For 30 Doradus we now know that the age for the stars is about 30 million years, which is young.'

Interestingly, the same cluster may have given birth to the progenitor star for Supernova 1987a, and Heap thinks it could be a site for a cluster of supernovae over the next million years, as these short-lived supergiants reach the end of their lives and die, spreading their elements out to be reused in future stars.

Betelgeuse

HST has turned its gaze toward another massive star in an effort to understand its variability. Betelgeuse (Alpha Orionis) is a red supergiant that marks the left shoulder of the winter constellation Orion the Hunter. FOC's image of the star is the first resolved image by any telescope of a star other than the Sun. The size of the star is immense. If Betelgeuse were placed at the center of the Solar System, its outer atmosphere would extend beyond Jupiter.

Figure 3.21. (a) The constellation Orion with Betelgeuse in the upper left shoulder. **(b)** FOC image of Betelgeuse shows that the star has a massive atmosphere, especially when scaled against the size of the orbits of Earth and Jupiter. See text for discussion. *(A. Dupree, Center for Astrophysics, NASA; ESA)*

(a)

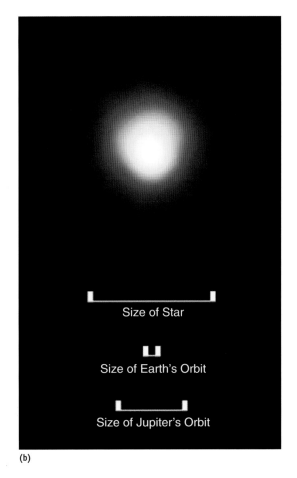

Size of Star

Size of Earth's Orbit

Size of Jupiter's Orbit

(b)

(a)

(b)

(c)

Figure 3.22. This is a NASA/HST image of the cool red giant variable star Mira, in the constellation Cetus. It is accompanied by its hot companion star. The top image **(a)** was taken on December 11, 1995, with the Faint Object Camera. Mira A is to the right, Mira B to the left. The stars are separated by 0.6 arcseconds (about 70 times the distance between the Earth and the Sun). The lower two panels are reconstructed images. **(b)** shows that Mira has an odd, asymmetrical shape. This could be caused by its expansion–contraction cycles, or is possibly due to the presence of unresolved surface spots. **(c)** In ultraviolet light, Mira appears to have a small hook-shaped appendage, possibly indicating the presence of material being drawn from Mira by the smaller star. It could also be material in Mira's upper atmosphere being heated through interaction with the companion star. *(Margarita Karovska, Harvard-Smithsonian Center for Astrophysics; NASA)*

A surprise is that the contours of brightness in the image are not circular. It appears that Betelgeuse has a bright, hot spot some ten times the diameter of the Earth and at least 2000 kelvin hotter than the rest of the surface. Since Betelgeuse is known to be a variable star, perhaps the movement of the hot spot across the face of this supergiant as it rotates could be the cause of brightness changes. Further study should help us understand the relationship between the spot and the star's variability.

Binary stars

Stars are often grouped in associations, in pairs or clusters, though some, like the Sun, go their way alone through the galaxy. Binary star systems are probably the most common grouping of stars, and they seem to be found everywhere. These are pairs of stars in mutual orbit around each other, and they have come in for a fair amount of study by the HST Astrometry Team. Studying how binary stars orbit each other is useful for determining masses of the two stars, and if astronomers can see them eclipse each other, they can confirm the radius of each star. This is one of the goals of astrometry.

HST is looking at many binary stars in an effort to determine their mutual interactions. Otto Franz of Lowell Observatory is part of the HST Astrometry Team making observations of binary stars. Although he and others have calculated binary orbits using earth-bound instruments, one drawback with these types of observations is that it is nearly impossible to resolve the two stars when they are closest to each other. The Astrometry Team uses HST to 'fill in' those parts of the orbit that cannot be seen with ground-based instruments.

This was a difficult project in the first years of HST's existence because of the pre-servicing mission problems with spherical aberration and jitter. Even before the repair, Franz said the data were pretty good: 'We had data with an accuracy of 1 milli-arcsecond per observation. That is pretty spectacular.'

One milli-arcsecond is a very small area of space. If you had that sort of resolving power with your eyes, for example, you would be able to spot a letter on this page from a distance of 200 kilometers.

Separating close binary stars has become easier in recent years. Omicron Ceti, also known as Mira, has a hot companion star. On December 11, 1995, the FOC observed these two stars and resolved the distance between the two. The stars, as shown in Figure 3.22, are separated by a distance that is 0.6 arcsecond. That is about 70 times the distance between the Earth and the Sun!

Further observations of Mira by HST showed that the star itself has an odd, asymmetrical shape. Since Mira is a well-known variable star, this shape may be tied to the changes the star goes through during the expansion–contraction cycle that manifests itself to us as changes in the star's brightness.

NGC 6624

Sometimes binary stars are so close together that the members interact gravitationally, and, in some cases, dump material on each other. HST has observed at least three incidents of this sort of 'stellar dumping'.

NGC 6624 is a globular cluster approximately 28 000 light years away from Earth in the direction of the constellation Sagittarius. Astronomers have known for years that this region is a strong source of x-ray bursts. Figure 3.23 shows ultraviolet and visible light images of the core of NGC 6624. The HST's Faint Object Camera isolated the source by searching for the ultraviolet light given off by a hot disk of gas surrounding the two stars.

The physical system is fascinating. One star is a white dwarf and the other is a neutron star. The white dwarf is distorted by the intense pull of the neutron star into a pear shape. The two stars are only about 160 000 kilometers apart (less than one-half of the distance between the Earth and the Moon), and they orbit around each other every 11 minutes. The gravitational pull constantly strips helium from the surface of the white dwarf, producing a steady fall of

material onto the neutron star, in turn producing a steady stream of x-rays. When large amounts of helium accumulate, nuclear fusion occurs to produce the very intense x-ray emissions that earned the name 'x-ray burster' for this pair.

Nova Cygni 1992

One type of binary star interaction is the sharing of material that produces what astronomers call a 'nova'. Novae occur when one star in a binary pair dumps material onto the surface of the white dwarf: what we see is basically a thermonuclear explosion. Nova Cygni 1992 was discovered from the ground, and imaged by HST soon after its latest eruption. The FOC imaged the ring of light formed on the rapidly ballooning bubble of gas coming from the star in 1993, and again right after the First Servicing Mission. The ring had expanded 35 billion kilometers from the star in the intervening seven months, giving scientists a good idea of how fast the gases are moving out from the site of the explosion. As the cloud of material expands from the star, it grows thinner and more tenuous, and we may see the circular ring grow more egg-shaped as the gas spreads out. If the Sun had a shell of gas that started this rapid rate of expansion in January, by early February it would have passed the orbit of Pluto. In July, the edge of the ring would have expanded to cover 1 per cent of the distance between the Sun and its closest stellar neighbor, Proxima Centauri.

As we can see, binaries exhibit a rich set of phenomena. We should remember that most stars are in binary systems, so studying different types of binary systems and the circumstances of their interactions provides a unique window to understanding many of the stars in the galaxy.

Star clusters

Stars can also be bound together gravitationally into associations called clusters. There are *open clusters*, which are generally thought of as loose-knit associations of stars that are found almost exclusively in the plane of the galaxy. Generally, they contain up to a few hundred stars, and, over time, the members of the cluster can escape the weak gravitational field that binds the cluster together. The Pleiades and Hyades in the constellation Taurus are good examples of open clusters.

Figure 3.23. Ultraviolet **(left)** and blue **(right)** FOC images of the core of the globular cluster NGC 6624. The ultraviolet image is dominated by a single source. In the blue image, this source, marked, is inconspicuous and is displaced from the center of the cluster. This source is a very hot star and x-ray source. *(Ivan King, University of California at Berkeley; NASA; ESA)*

Figure 3.24. (Above) FOC image of Nova Cygni 1992, before COSTAR (May 31, 1993) and after COSTAR (January 1994, **top right**). The images show the expansion of the ring of ejected material and the improved resolution of the FOC with COSTAR. See the text for further discussion. *(F. Paresce and R. Jedrzejewski, STScI; NASA; ESA)*

Figure 3.25. M15 (NGC 7078), the globular star cluster in Pegasus. *(D. DiCicco, Sky & Telescope Magazine)*

Figure 3.26. (Opposite) FOC image of faint blue stars at the core of M15. Fifteen very hot, faint stars are shown in this ultraviolet image, which is about 2 light years across. Most of them are concentrated within 0.5 light year of the center of the core. A likely explanation for their appearance is that the crowded conditions at the center of M15 produced close encounters which resulted in the outer layers of the stars being stripped off, leaving the 'naked cores'. *(G. de Marchi, STScI and the University of Florence, Italy; F. Paresce, STScI; NASA; ESA)*

Sometimes when you scan the sky at night with binoculars, you might view tightly packed associations of stars called *globular clusters*. There are more than 146 of them associated with the Milky Way. Globular clusters typically contain tens or hundreds of thousands of very old stars, bound together in a tight gravitational ball from which there is no escape. The night sky from anywhere inside a globular cluster would blaze with the light of a hundred thousand suns!

M15

One of the chief difficulties with observing globular clusters is in distinguishing individual stars, particularly when they are tightly packed together in the core of the cluster. Despite the density of the starfields in globulars, HST has had good success in resolving single stars in very

crowded fields. Figure 3.25 shows the globular cluster M15 located in the constellation Pegasus, some 30 000 light years away from Earth. To an observer using a pair of binoculars, it is visible as a hazy-looking patch of light about one-third of the Moon's diameter. Through a small telescope it begins to look more like a shimmering globe of stars. This cluster has long been known to be home to a large number of variable stars. HST has imaged the central region of M15, and Figure 3.26 shows HST's view – a region about 2 light years across. In the central core, 15 very hot, blue stars are clearly visible. These stars pose a mystery, since hot blue stars are generally thought to be young stars, and globular clusters are not usually known for their young star populations. Indeed, the historical view of globulars is that they consist of stars that may be as much as 13 to 15 billion years old! Why would a globular cluster contain what appears to be hot young stars? One good explanation for these blue stars is that they are 'naked cores' of old stars – that is, the central portions of stars that have been stripped of their outer envelopes of gases by collisions with each other. This type of interaction can only take place where the stars are so crowded together that the chance for very close encounter (which gravitationally pulls material from both stars) is exceptionally good.

According to astronomer Guido De Marchi, the find was amazing: 'These objects represent a totally new population of very blue stars. When we started wondering what they could be, we realized that they may be among the first observed stars to have been stripped of their atmospheres.'

This result lends support to the idea that the evolution of stars can be changed drastically by close encounters with other stars. A more recent image of this cluster is shown below.

Another globular cluster under study – 47 Tucanae – also has a good-sized population of massive blue stars. These were first imaged in late 1990 by the FOC. (see Figure 1.16, page 20)

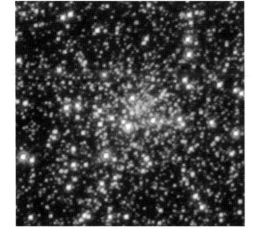

Figure 3.27. The densely populated heart of globular cluster M15. The WF/PC-2 centered its planetary camera over the core of the galaxy, looking for evidence of either a black hole or a 'core collapse' driven by the intense gravitational pull of the stars. The larger picture shows the 28 light-year-wide central portion of M15, photographed with the planetary camera. The cluster's central region is shown in the smaller image, showcasing a 1.6 light-year-wide region. Hot stars appear blue, while cooler stars appear reddish-orange. *(P. Guhathakurta, UCO/Lick Observatory, UC Santa Cruz; B. Yanny, Fermi National Accelerator Lab; D. Schneider, Pennsylvania State Univ.; J. Bahcall, Inst. for Advanced Study;NASA)*

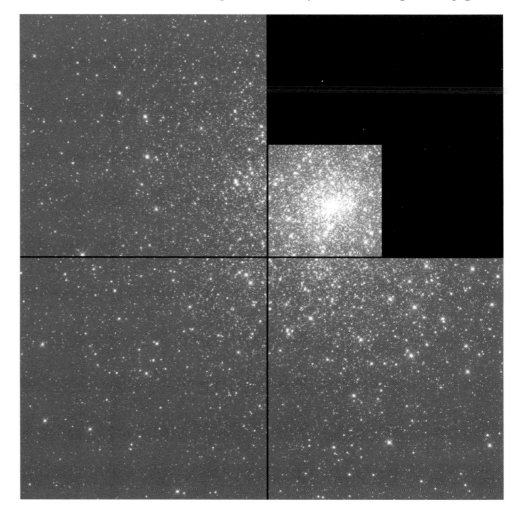

Scientists readily spotted a collection of 'blue straggler' stars in the crowded confines of the cluster. These are massive stars that have passed into old age, but have somehow managed to rejuvenate themselves back to a hotter and brighter state, most likely by colliding with each other and merging. Because these stars appear to be so massive, it is also possible that they may actually be double star systems. If so, their action may influence the motions of other stars within the cluster.

Chemically peculiar stars

Stars do not have to flaunt their naked cores to catch an astronomer's interest. Sometimes their chemical compositions send out a coy 'come look at me' call. Recall chemical abundances from our discussion of the interstellar medium. Many stars have abundances of chemical elements that are close enough to the standard cosmic abundances that astronomers call them 'normal'. Sometimes, however, astronomers find a star where the observed abundances are so abnormal – so far from the so-called normal abundances – that they call out for special study. When these kinds of stars are studied in depth, the following questions come to mind. Do we really understand the nucleosynthesis of the elements or is it possible that we are observing a special case? Is the chemical makeup of the region on the star that we are observing typical of the entire star? This is an important question, because one of the keys to understanding the star lies in determining its chemical makeup. If elements and their isotopes are stratified on the star – that is, if they only show up in particular layers of the star – it would be a mistake to assume that those elements exist throughout the star.

Astronomers call stars with unusual chemical abundances 'chemically peculiar' stars. GHRS Co-Investigator Dave Leckrone made observations of Chi Lupi, a chemically peculiar star with a surface temperature of about 10 000 kelvin, and a small binary companion. A section of Chi Lupi's spectrum is shown in Figure 3.28. Leckrone described these observations as an attempt to construct the most detailed road map of the spectrum of a star anyone had ever done in the ultraviolet. 'What we're trying to do with Chi Lupi is to define a pathfinder study,' he said. 'Obviously, we would like to observe many different stars with many different kinds of peculiarities, covering a range of temperature, gravity, magnetic and non-magnetic characteristics, and so on. We want to look at the basic 'zoology' of these stars and see what determines some of these peculiarities we see.'

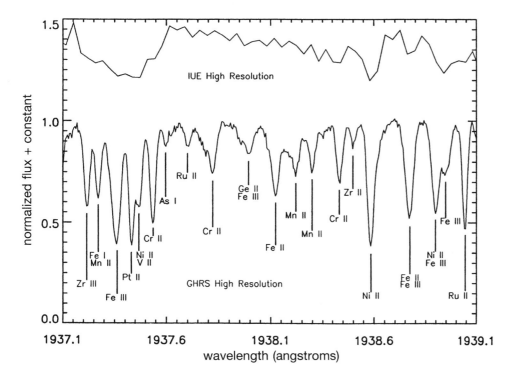

Figure 3.28. A 2-angstrom segment of the Chi Lupi spectrum taken with the highest resolution mode of the GHRS. The lines in the spectrum are cleanly resolved, whereas they were not in the IUE spectrum (above). The elements identified are: arsenic (As); chromium (Cr); iron (Fe); germanium (Ge); manganese (Mn); nickel (Ni); platinum (Pt); ruthenium (Ru); and zirconium (Z). *(David Leckrone, NASA-Goddard Space Flight Center; Glenn Wahlgren, Computer Sciences Corporation and GHRS Science Team)*

When Leckrone and his team started their study of Chi Lupi, only a dozen or so elements had been seen spectroscopically in the star's atmosphere. 'The number is up to 49 elements, with 67 different ionization stages,' Leckrone said. 'We're planning on producing a team project Chi Lupi atlas of elements, and in going through the drawings for this project, we've found two or three more elements waiting be analyzed. We're going to be well over 50 elements before this is all over.'

For chemically-peculiar stars, the biggest challenge to understanding these abundances, is to figure out *why* they are peculiar. Chi Lupi certainly fills the bill. 'This is a star with really bizarre abundances, including a lot of heavy metals that are enhanced,' Leckrone said, 'and you see a lot of detail about the individual elements' spectra that you wouldn't see in other stars where these elements are not so high in abundance.'

Leckrone thinks the peculiarity is something confined to the surface layers of the stars. The element mercury (Hg) beautifully illustrates some of the bizarre results. Figure 3.29 shows the observations of a mercury line and the best-fit theoretical profile.

Two results come out of this analysis. First, the mercury seems to be in the form of mercury 204. If you buy a mercury thermometer at a hardware store, the mercury is a mixture of the mercury isotopes 198, 199, 200, 201, 202 (no 203), and 204. The fit shown is for 99 per cent mercury 204, the heaviest stable mercury isotope.

Secondly, the abundance value is very high, some 100 000 times the value in the Solar System. Similar very high values are found for platinum (Pt) and gold (Au).

It is very likely that Chi Lupi shows such a high amount of these heavy elements in its spectrum because we are only seeing the top layers of the star. Chi Lupi's photosphere may itself be layered, which means that specific elements have congregated in specific places and we just happen to be examining light emitted from those places. If that has not happened, then we must consider other explanations, such as some unusual nuclear processes running amok deep within the star. So far, the models to explain this abundance and the processes that produce it within the star are not complete.

A good possibility is that the atmosphere of the star is unusually stable. According to Leckrone, the Chi Lupi studies have allowed his team to construct models of the very subtle physical processes that occur in the evolution of stars. Chi Lupi is, itself, an unusual case

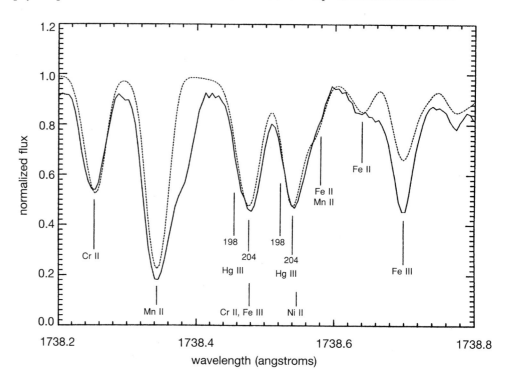

Figure 3.29. A 0.6-angstrom segment of the Chi Lupi spectrum (solid line) taken with the highest resolution mode of the GHRS. In addition to lines already identified in Figure 3.28, the segment shows two lines of mercury (Hg). The position of the lines changes with the particular isotope present, and the positions for the isotopes 198 and 204 are marked. The spectrum clearly matches isotope 204 (dotted) and is not close for 198. The analysis indicates that the mercury is at least 99 per cent isotopic. *(David Leckrone, NASA-Goddard Space Flight Center; Glenn Wahlgren, Computer Sciences Corporation and GHRS Science Team)*

because of the elements it contains. According to Leckrone, Chi Lupi has no rotational velocity in space: 'It's extraordinarily slow and the question you have to ask is: what happened to it? Was that just a pure accident of its birth, or is it because of interactions with its binary companion?'

Chi Lupi is an unusually quiescent star, and Leckrone suggests that this may explain the effects he has measured. 'The star itself is unusually quiescent. You have a very stable situation in the atmosphere of this star, in its inner envelope of gases,' Leckrone explains. 'Gravity is trying to tug atoms and ions downward and radiation pressure is trying to push them upward and there's this competition between the two. A diffusion process goes on, with material being pumped back and forth, but we're not sure it's in equilibrium.'

Small differences in radiation pressure – the pressure exerted by light streaming away from the core of the star – could concentrate isotopes and elements in regions of the star under observation. Understanding the concentrations of elements and the diffusion process that seems to drive the star's atmosphere makes Chi Lupi a test-bed for the understanding of hydrodynamical processes that affect stellar evolution. Leckrone's study of Chi Lupi remains very much a work in progress, and he hopes to apply this work to the study of other chemically-peculiar stars with HST.

One interesting side effect of Leckrone's Chi Lupi work is the impact it has had on atomic physics. Understanding the spectra as shown in Figures 3.28 and 3.29 requires atomic physics data of higher quality than is sometimes available to fit the strength and location of the observed lines. This work has led to a unique collaboration between atomic physicists – who try to pinpoint elemental species in the laboratory setting – and the astrophysicists, who observe these species in space. In the case of Chi Lupi – as with the data from observations of the interstellar medium – Leckrone and Swedish astrophysicist Sveneric Johansson of the University of Lund are 'filling in the table of the elements' using these observations of naturally occurring elements in space. It is important to remember that these advances could not have taken place before an operational HST and GHRS, and Figure 3.28 shows a comparison of the observations of Chi Lupi made by HST and IUE. Recognizing the unique value of this work, the Swedish Academy of Sciences awarded Johansson the prestigious Wallmark prize for his achievements, citing his research in the border field between atomic physics and astronomy, particularly high-resolution spectroscopy using the Hubble Space Telescope.

The prize was presented to Johanssen by His Royal Highness King Carl XVI Gustaf of Sweden on March 31, 1995. The honors for this work have continued, with Leckrone receiving an honorary degree from the University of Lund in 1995.

Faint stars and dark matter

One of the most pressing issues in modern astronomy is the question of so-called 'dark matter'. Simply put, there is overwhelming evidence that approximately 90 per cent of the mass of the universe is invisible to us because it does not emit any radiation or that radiation is so feeble we cannot detect it. There are many kinds of 'dark matter', some quite exotic (as we will discuss in the context of cosmology in Chapter 5). Here we address the issue of faint stars being part of the as-yet unseen and unmeasured mass of dark matter in the universe. Hubble investigations have uncovered some very faint stars. Figure 3.32 shows a diminutive companion to the dwarf star GL 105A. The companion (upper right) is called GL 105C and is 25 000 times fainter than GL 105A in visible light. The spikes and the white bar coming from GL 105A are artifacts produced by the telescope and the WF/PC-2 when a very bright object is in the field of view. The HST image was taken in PC mode and confirms the discovery by David Golimowski at Palomar Observatory.

It is useful to introduce some terminology for this field at this point. Stars with masses lower than the Sun, but that are still capable of hydrogen burning, are called red dwarfs. These can occur for masses down to about 8 or 9 per cent of the Sun's mass. Stars with lower masses cannot burn hydrogen and are called brown dwarfs. The mass estimated for GL 105C is close to the red dwarf/brown dwarf dividing value.

Figure 3.30. King Carl XVI Gustaf of Sweden, presenting Sveneric Johansson with honors for his research into Chi Lupi, March 31, 1995. (*Courtesy Lars Falck, Stockholm*)

Figure 3.31. David Leckrone (center, with laurel wreath), at Lund University on May 31, 1996. He is preceded by a marshall (red bow) of the University procession. (*G. Wahlgren, Computer Sciences Corporation*)

Figure 3.32. WF/PC-2 image of the star Gliese (GL) 105A, lower left, and its companion Gliese 105C, at upper right. *(D. Golimowski, Johns Hopkins University; NASA)*

Figure 3.33. A suspected brown dwarf orbiting the star Gliese 229B. On the left is the Palomar Observatory discovery photograph, taken October 27, 1994. The right image is the HST's WF/PC-2 image, taken November 17, 1995. *(T. Nakajima, S. Kulkarni, California Institute of Technology; S. Durrance, D. Golimowski, Johns Hopkins University; NASA)*

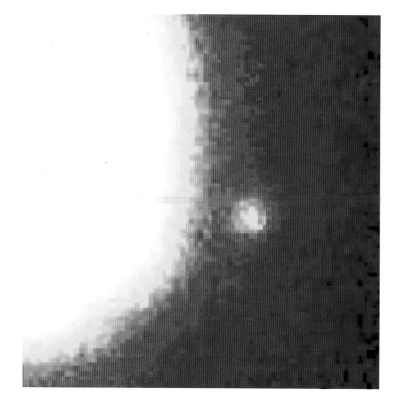

Figure 3.34. (a) (Opposite top) The search for red dwarfs in the globular cluster NGC 6397. The left image is a ground-based view. The WF/PC-2 field of view is marked in the exploded box.
(b) (Opposite bottom) A comparison of the HST image of this cluster with a simulated image, assuming that faint red dwarf stars (marked in diamonds) were abundant in the Milky Way Galaxy. *(F. Paresce, STScI; NASA; ESA)*

An unquestionable detection of a brown dwarf is shown in Figure 3.33. This object, GL 229B, orbits the red dwarf star GL 229. The figure shows the ground-based image from Palomar Observatory and the WF/PC-2 image. In the HST view, light flooding produces the spike from the telescope optics.

The mass of GL 229B is estimated to be 20 to 50 times the mass of Jupiter. The distance between GL 229 and GL 229B is roughly the same as the distance between Pluto and our Sun. By definition, then, the object GL 229B is certainly a brown dwarf.

These discoveries show that very faint stars exist. However, these are stars in orbit around brighter stars. Such stars cannot make a major contribution to the dark matter problem. For this to happen, we would need to find large numbers of these objects in star clusters or in the halo of our galaxy. So far, the results are not encouraging.

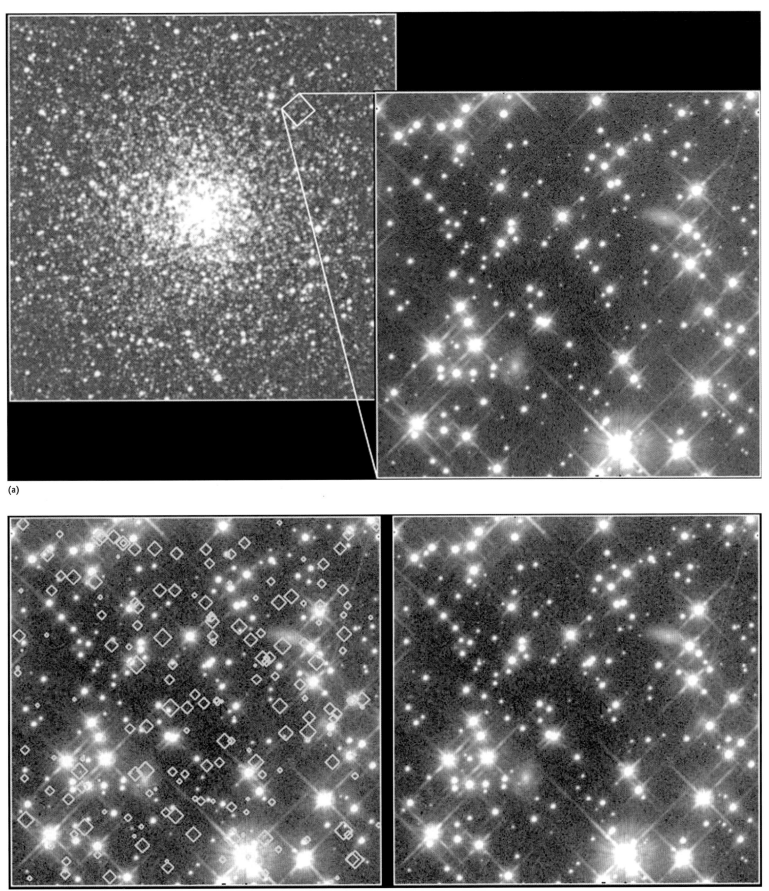

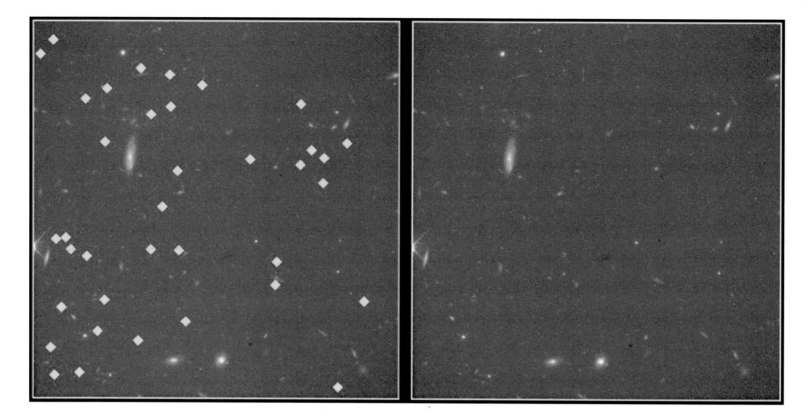

Figure 3.35. A search for faint stars in the galactic halo. If dark matter was all locked up in faint red stars, there would be 38 of them in the image (as represented by the diamonds added to the left image). Comparison with the real image on the right shows a lack of such stars. *(J. Bahcall, Institute for Advanced Study, Princeton; NASA)*

Figure 3.34 shows images of the globular cluster NGC 6397. A simulated image is shown as if faint red dwarf stars (diamonds) were abundant. The image shows far fewer stars than expected, and background galaxies can be seen through the cluster field. Stars one-fifth the mass of the Sun are abundant, and for every star with the mass of the Sun, there are about 100 stars with one-fifth the solar mass. Stars smaller than one-fifth of a solar mass are rare, and some astronomers are skeptical about their existence. Francesco Paresce of the Space Telescope Science Institute and European Space Agency, and leader of the team investigating NGC 6397, has noted: 'The very small stars simply don't exist.'

This attitude does not mean that astronomers are giving up the search for these objects. An HST image of a patch of sky has been used to search for faint red stars in the halo of our galaxy. Figure 3.35 shows the type of image expected if the faint red stars were abundant (with simulated stars) and the HST image. The HST WF/PC-2 image is so devoid of stars that distant galaxies are easily seen. The region was chosen to minimize the contribution of faint stars near the plane of the Galaxy. Thus, the possibility that red dwarf stars could be an important source of dark matter appears to be ruled out. John Bahcall, Institute for Advanced Study, Princeton, and leader of the team searching for the faint halo stars, has remarked: 'Our results increase the mystery of the missing mass. They rule out a popular but conservative interpretation of dark matter.'

Obviously, we need to examine other candidates for the dark matter, and we will explore those candidates in the discussion of cosmology in Chapter 5.

Planetary nebulae

When stars about the size of Earth reach their final stage of evolution, they become planetary nebulae (so named because they appear in telescope views as stars with surrounding rings of gas and dust). What is happening inside the star is quite interesting. Nuclear fusion in the core has ceased, and the core is a degenerate carbon–oxygen melange about half the mass of the Sun. Surrounding the core are hydrogen- and helium-burning shells. The whole thing is wrapped in a hydrogen envelope, and stretching away from the star is a huge shell of gas blown off during mass loss. The star continues to produce energy, and burns off the outer layers. This process exposes the hot core, and the star's surface temperature increases. Eventually it reaches a temperature of about 25 000 Kelvin, by which time it is hot enough to produce ultraviolet radiation, which ionizes the shell and lights it up, creating a thing of ghostly beauty. The glowing shell is like a fossil record of the end stages of the dying star within.

Etched Hourglass Nebula

For years, astronomers carried around in their heads an image of the 'typical' planetary nebula, based largely on the appearance of the best-known example: the Ring Nebula in Lyra. It fitted the theory perfectly, and perhaps satisfied some inner need for symmetry. As better and better telescopes began to find other planetary nebulae, this picture started to change. As part of its long-term study of stellar evolution, HST has peered into the heart of several not-so-symmetrical nebulae, revealing a lovely variety of shapes and asymmetries.

The WF/PC-2 image of one planetary nebula called MyCn 18 reveals an hourglass shape with an intricate pattern or etchings in its walls. Light from ionized nitrogen is rendered in red, from hydrogen in green, and from doubly-ionized oxygen in blue. These details could not be seen previously in ground-based images.

The complexity and asymmetry of this nebula is challenging. What produces the overall shape? And why is the 'central star', the white dot *near* the center, not *at* the center? Speculations on the source of the structure range from beams of matter emanating from the central star and impinging on the hourglass walls, to an unseen stellar companion disturbing the shape through gravitational effects. For now, the long-standing puzzle of how some planetary nebulae acquire their complex structure remains alive and well.

Figure 3.36. The Ring Nebula, M57. This famous planetary nebula is the aftermath of the death of a star like the sun. *(J. Newton)*

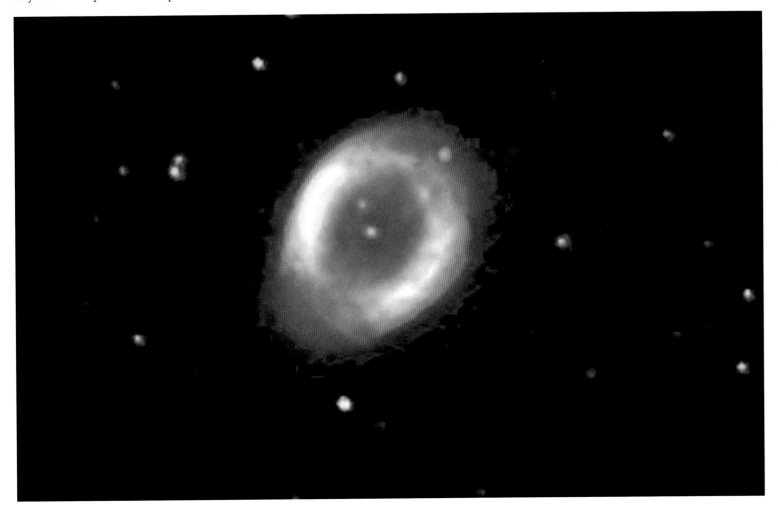

Figure 3.37 This WF/PC-2 image of the Etched Hourglass Nebula, MyCn 18, shows some of the beautiful and intricate detail seen in planetary nebulae. *(R. Sahai and J. Trauger, Jet Propulsion Laboratory; WF/PC-2 Science Team; NASA)*

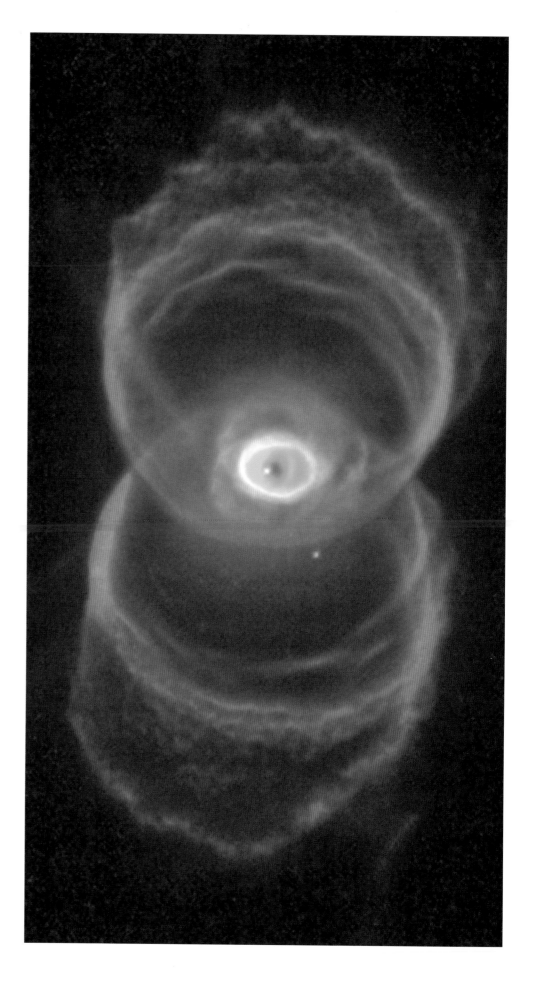

Stars and the interstellar medium

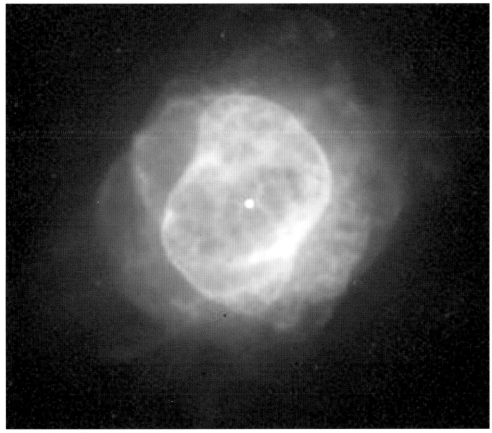

NGC 6884 *(Howard Bond, STScI; NASA)*

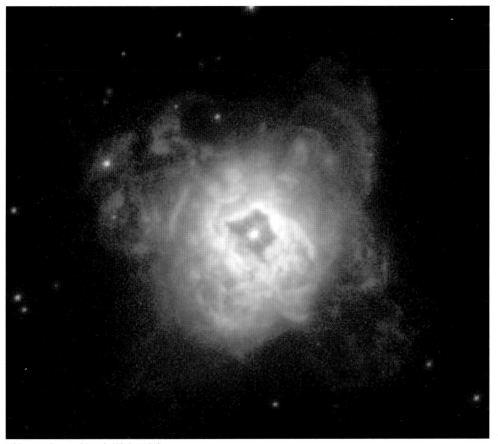

NGC 5315 *(Howard Bond, STScI; NASA)*

HST and the stellar zoo

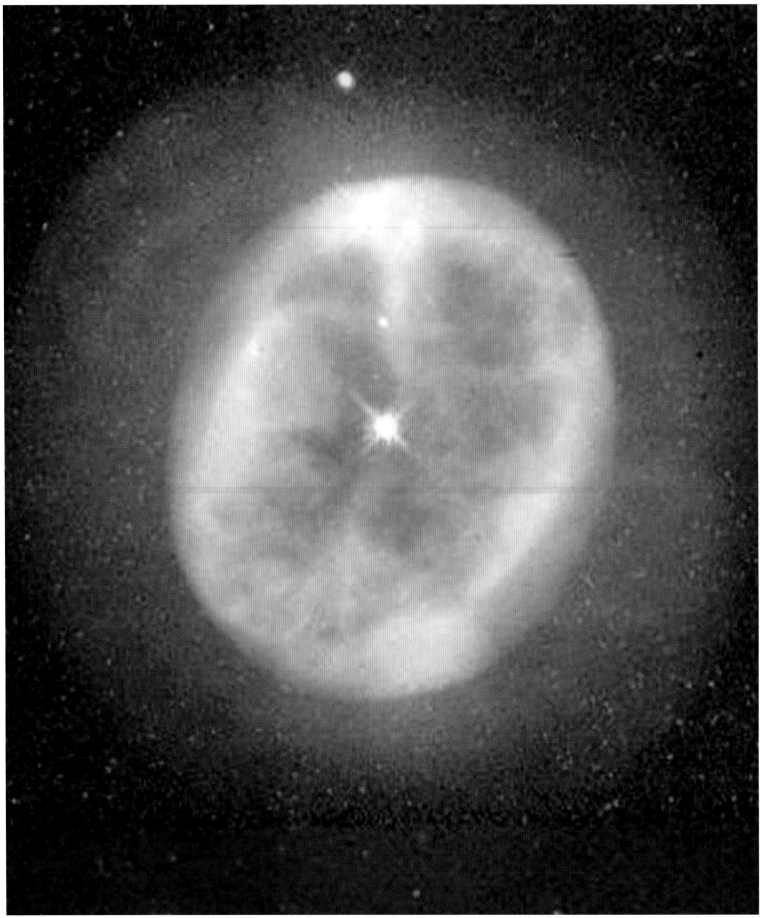

NGC 2022 *(Howard Bond, STScI; NASA)*

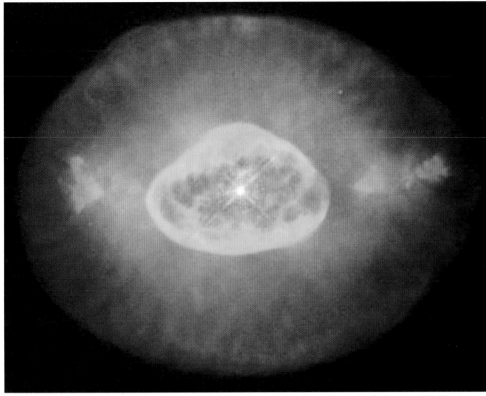

NGC 6826, also known as 'The Blinking Planetary Nebula'. *(Bruce Balick, Jason Alexander, University of Washington; Arsen Hajian, U.S. Naval Observatory; Yervant Terzian, Cornell University; Mario Perinotto, University of Florence (Italy); Patrizio Patriarci, Arcetri Observatory; Howard Bond, STScI; NASA)*

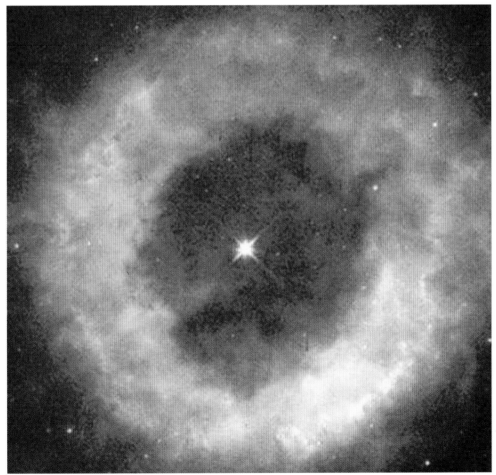

NGC 6369, nicknamed 'The Little Ghost Nebula'. *(Howard Bond, STScI; NASA)*

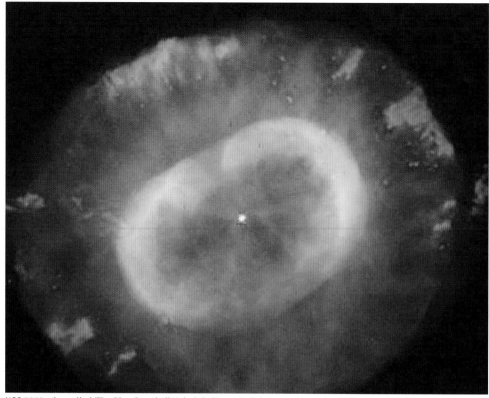

NGC 7662, also called 'The Blue Snowball Nebula'. *(Bruce Balick, Jason Alexander, University of Washington; Arsen Hajian, U.S. Naval Observatory; Yervant Terzian, Cornell University; Mario Perinotto, University of Florence (Italy); Patrizio Patriarci, Arcetri Observatory; Howard Bond, STScI; NASA)*

NGC 6578 *(Howard Bond, STScI; NASA)*

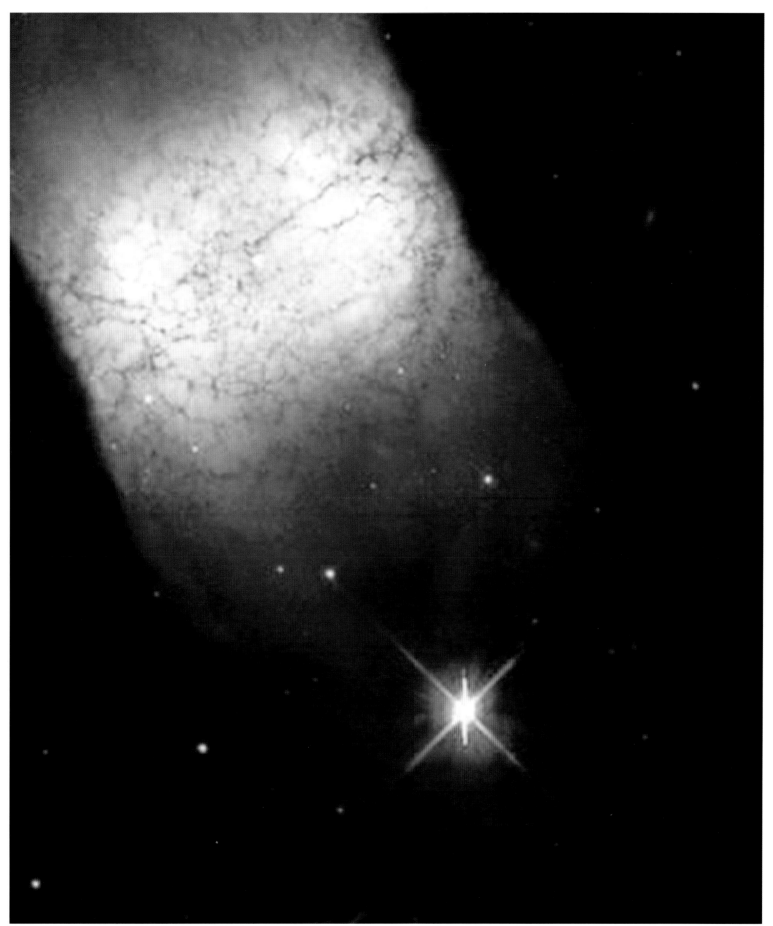

IC 4406, called 'The Bipolar Nebula'. *(Howard Bond, STScI; NASA)*

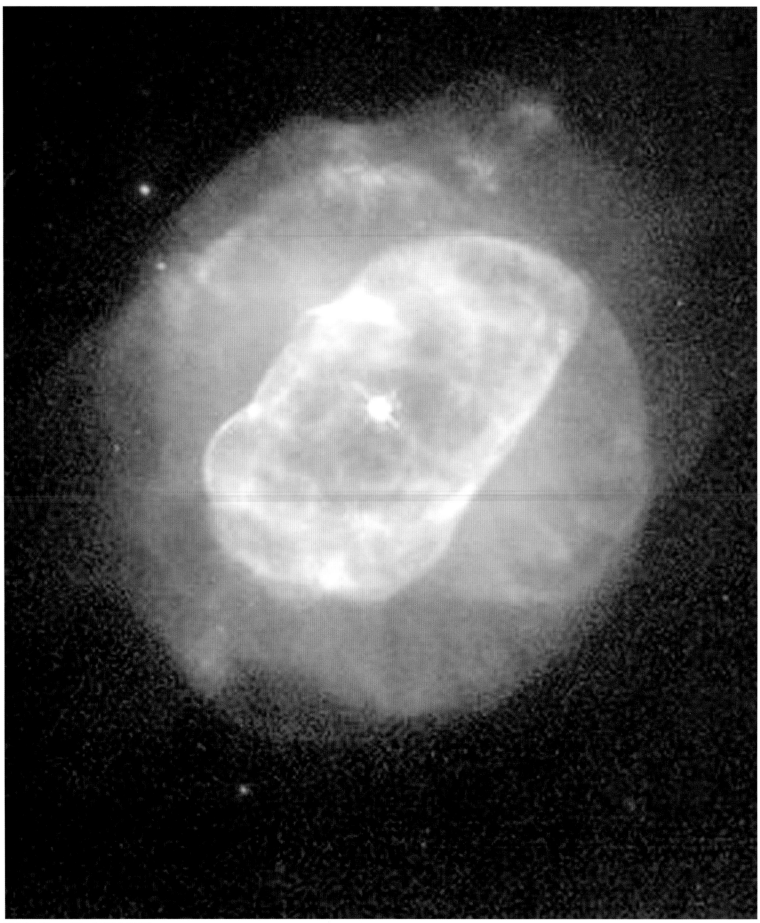

IC 5882 *(Howard Bond, STScI; NASA)*

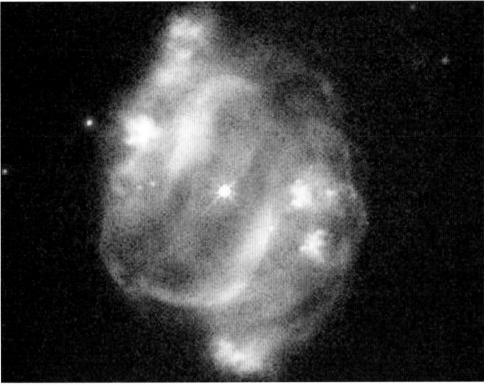

NGC 5307, 'The Christmas Tree' *(Howard Bond, STScI; NASA)*

AFGL 915, 'The Red Rectangle' *(Howard Bond, STScI; NASA)*

Hubble 5 *(Bruce Balick, University of Washington; Vincent Icke, Leiden University (Netherlands); Garrelt Mellema, Stockholm University; NASA)*

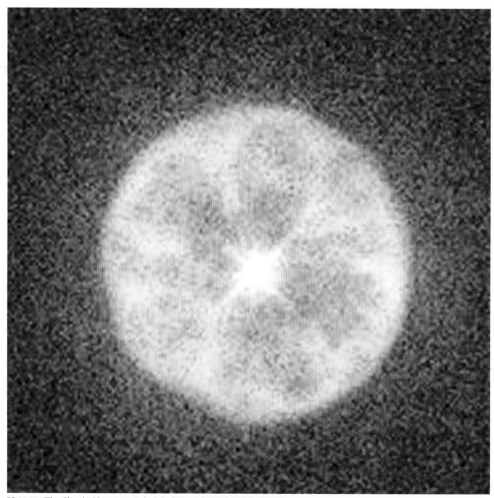

IC 3568, 'The Simple Planetary Nebula'. *(Howard Bond, STScI; NASA)*

Cat's-Eye Nebula

One of the most complex planetary nebulae observed, NGC 6543, lies in the constellation Draco. Nicknamed the Cat's-Eye Nebula, it is estimated to be at least 1000 years old. Seen through WF/PC-2 in September 1994, the structure of the nebula shows jets of high-speed gas, concentric shells, and shock-induced knots of gas. At least one theory suggests that the star at the center might be a binary system. These dying binaries probably spent their last millennia losing mass through stellar winds. Because of their motion around each other, the shells of gas were twisted into complicated shapes. A fast stellar wind blowing off the central stars might have created the curious ellipsoidal shape of the inner gas shell. Surrounding it are two

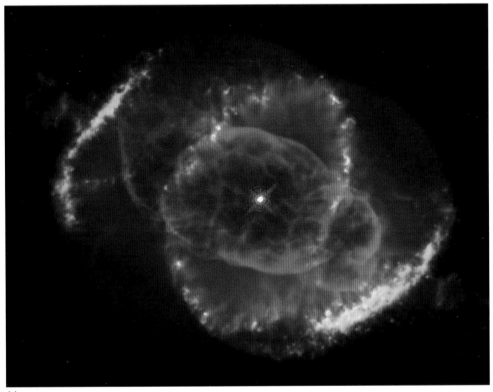

(a)

Figure 3.39. (a) WF/PC-2 image of NGC 6543, the Cat's-Eye Nebula. This planetary nebula lies 3000 light years away from Earth. **(b)** A schematic for the Cat's-Eye Nebula. *(STScI; NASA)*

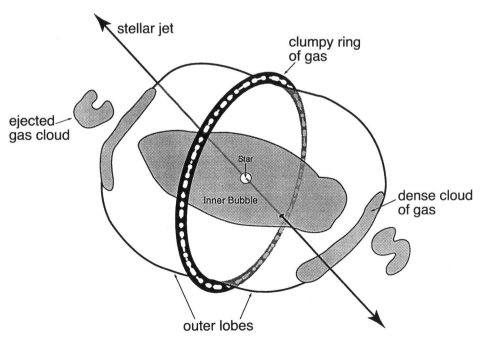

(b)

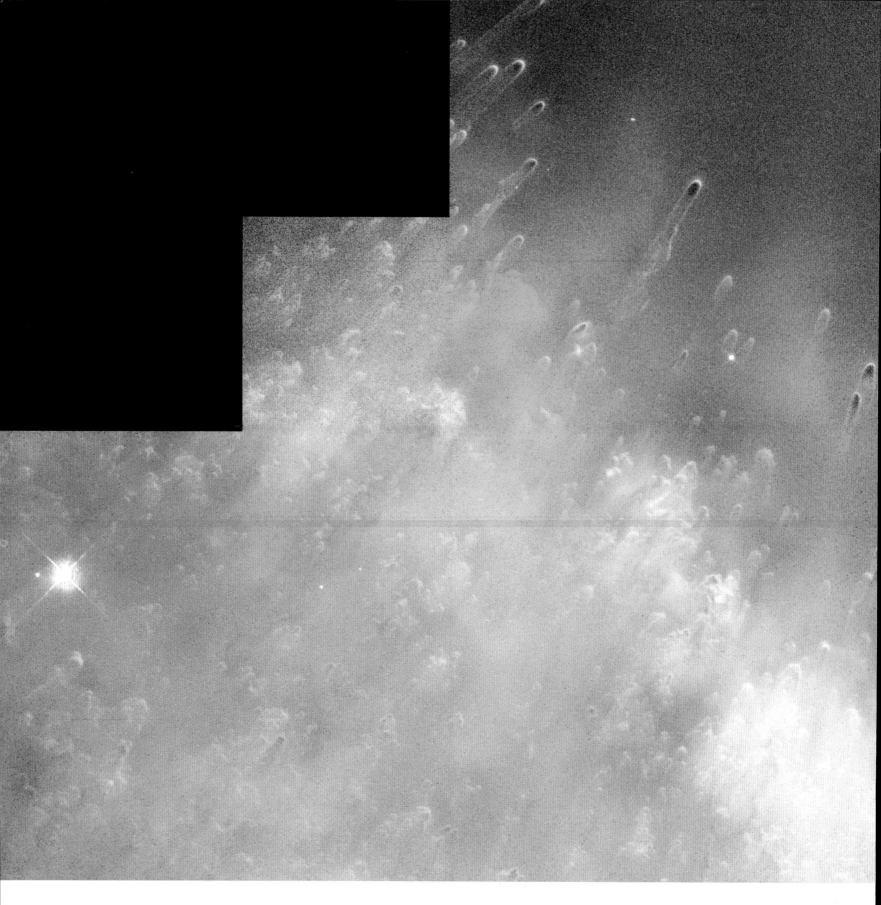

Figure 3.40. WF/PC-2 image of huge 'knots' of material in the Helix Nebula, NGC 7293. *(C.R. O'Dell and Kerry P. Handron, Rice University; NASA)*

larger lobes of gas blown away from the star. The bright arcs and curl-shaped structures might be formed by jets of gas. These jets seem to point in different directions from some of the features they have highlighted, suggesting they are wobbling as the stars orbit each other, possibly turning on and off like a beacon. To help understand the dynamics of the system, astronomers mapped the glowing gas clouds at wavelengths specific to particular colors. In Figure 3.39(a) red is hydrogen, blue denotes neutral oxygen, and green is ionized nitrogen.

Helix Nebula

HST studies of planetary nebulae continue to produce some of the most spectacular images in all of the HST collection. Witness the image of the Helix Nebula, the closest planetary nebula to Earth, in the constellation Aquarius. The condensations are called 'cometary knots'. Each giant 'comet head' is at least twice the size of our Solar System and each 'tail' stretches more than 100 billion kilometers. The WF/PC-2 recorded thousands of these knots.

The doomed, central star of the planetary nebula spews out hot gas which collides with the cooler gas surrounding the star. The cooler gas was probably ejected by the star around 10 000 years ago. The collision between the gases fragments the smooth cooler gas into the smaller 'droplets'. These droplets will eventually dissipate into interstellar space. For a specific planetary nebula, these objects can be seen as existing as part of a transitory phase. However, it may be that they are more common in our galaxy, and this will be verified if we find many planetary nebulae exhibiting this activity. Searches for similar features in other, more distant, planetary nebulae will be carried out by HST, and O'Dell hopes to re-image the Helix in a few years to search for motion of the knots.

Stellar death throes

Beyond giants and supergiants, planetary nebulae and chemically peculiar stars, astronomers reach into the realm of the outlandish stars – weird beasts like white dwarfs, neutron stars, and the ever-popular black holes. These intriguing members of the 'stellar zoo' are extreme examples of stars that are 'doing something'. What they are doing is dying, and that activity can range from the quiet fading away of a white dwarf to the violent explosion of a supergiant star as a supernova.

White dwarf stars

White dwarfs are faint old stars with masses less than $1.4M_S$. Most of them are about the same size as the Earth, but they are 300 000 times its mass. Some may have started life as stars a few times the size of the Sun, but blew off much of their material into surrounding space in a gentle stellar wind. Others had masses below $1.4M_S$. At the white dwarf stage of their lives, the nuclear furnaces inside these stars have stopped burning. They are dimming down, and at some point they will stop contracting. This happens when the electrons in the atoms of gas that remain in their cores push against the contraction and the star enters a state of equilibrium – no more contraction, no more energy source. What is left is a stellar mass so dense that a small spoonful of the star's material would weigh 10 tons! As an important end-point of stellar evolution, white dwarfs are a natural target for HST scientists, especially if they want to take spectra of those stars as they die. Because the gravity for white dwarfs is very high, the spectral lines are usually broad, and so the Faint Object Spectrograph, with its high sensitivity and ultraviolet coverage in ultraviolet wavelengths, was the instrument of choice.

If you study a white dwarf star, you should not expect to see much wind (if any) blowing away from it. If you have a spectrograph able to resolve the light from such a dim, distant object, you might be able to form an idea of which elements still exist in the leftover atmosphere of the star.

HST observations of the hot (effective temperature of about 62 000 kelvin) white dwarf G191.B2B have revealed the presence of highly ionized metals in the star. Ionized metals

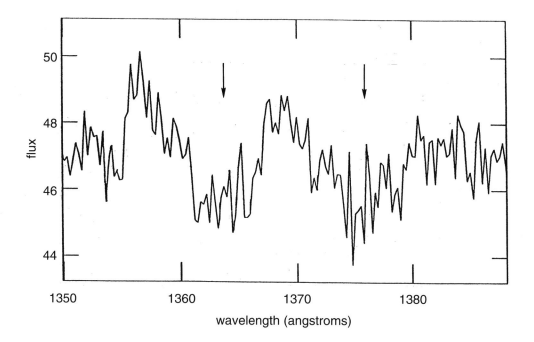

Figure 3.41. FOS spectrum of the white dwarf G191.B2B. Two of the strongest iron (Fe V) lines are marked by arrows. See text for discussion. *(Edward M. Sion, Villanova University)*

betray themselves through their spectra. In Figure 3.41, two broad absorption features at 1364 angstroms and 1376 angstroms are produced by iron ionized four times (Fe V). While Fe V lines are the most numerous, absorption lines of other elements show up too.

Why should hot iron be so important here? Unless there is some opposing mechanism, heavy elements like iron should sink into the white dwarf and be invisible in a spectrum. If we still see these elements in the spectrum, then something is happening to levitate or even expel these elements from the star. It is possible that high radiation pressure from within the star is causing this levitation effect. Maybe there are dense shells of material surrounding the star, and these shells are showing off their ionized iron content. Perhaps mass loss from white dwarfs actually occurs. If so, it challenges the conventional understanding of these slowly dying hot stars.

Supernova 1987a

Nothing catches the attention of astronomers quite like the final, cataclysmic explosion of a supergiant star called a supernova. These temporary but spectacular outbursts can be unbelievably bright – sometimes rivaling the luminosity of entire galaxies. For that reason, supernovae are often used as standard candles to determine the distances to the galaxies they inhabit.

Because of their pivotal role in stellar evolution – from enrichment of the interstellar medium in heavy elements, creation of elements heavier than iron, production of neutron stars and black holes, and their probable role in the conception of new stars – supernovae afford astronomers who use HST a welcome research opportunity. Supernova 1987a (SN 1987a) in the Large Magellanic Cloud presented a spectacular chance to watch the aftermath of just such a stellar explosion. Since it was first observed by HST in August 1990, studies have focused on the cloud of debris rushing out from the supernova.

The first image from HST showed an elliptical, luminescent ring surrounding SN 1987a. This ring was seen glowing very faintly in previous ground-based and space observations, but the FOC image supplied a much improved view.

The ring was too far away from the supernova to consist solely of material that had been blown off during the event. The material was already in the neighborhood, probably as the result of massive, on going stellar wind activity by the progenitor blue supergiant star. The outgassed cloud was brightened by the light flash from the supernova, which reached the ring 240 days after the explosion.

What will happen next with SN 1987a? A cloud of stellar debris is rushing out from the center of the explosion, and, according to University of Colorado astronomer Dick McCray, calculating the exact time at which the expanding debris will contact the outer ring of material is an interesting problem. He is heavily involved in theoretical modeling of the scenario unfolding at SN 1987a. 'The bottom line is that it's going to impact the ring,' he says. 'Anywhere up to 2002 is when it's going to hit. We'll learn what that ring is. It's going to sparkle, it's going to have line profiles, it's going to be seen in x-rays as well as with HST.'

McCray believes that the ring is a unique phenomenon. In some ways it is still a mystery, but he has his own theory about how it formed: 'Everybody's been saying that the ring was ejected by the progenitor star, the supernova – but this new idea is that the ring was not ejected, but was a protostellar disk that was formed at the same time as the supernova progenitor was formed. When the supernova progenitor became a blue star, it eroded away this disk –

Figure 3.42. Supernova 1987a (the bright star at bottom) in the Large Magellanic Cloud on March 2,1987, as photographed with the CTIO Curtis Schmidt telescope by M. Bass. *(National Optical Astronomy Observatories)*

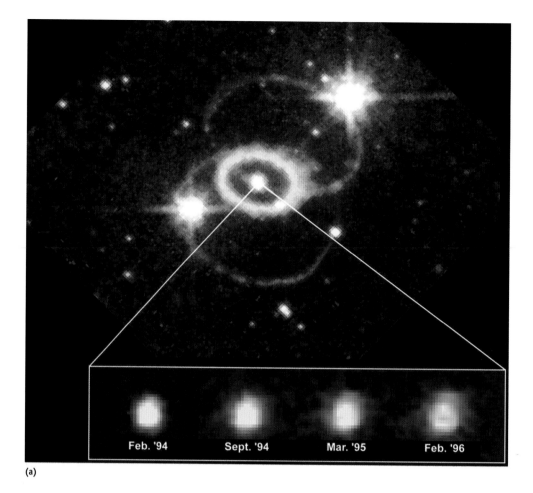

(a)

ate a big hole in the middle – and the outer part of the disk is still there and the ring is the inner rim. When the supernova went off, it made a flash and the flash ionized the inner rim of this disk and made it fluoresce. Space Telescope will make it absolutely clear whether it was a protostellar disk or was ejected.'

Supernova 1987a has not yet finished astounding astronomers. An even more spectacular image taken later (Figure 3.43a) shows the same ring at the center of the supernova site, plus a double set of rings flanking the supernova, a distinct change in appearance that has had McCray and other astronomers stumped for a workable explanation. Chris Burrows, at the Space Telescope Science Institute, who was one proponent of the compressed gas shells theory for the central 'celestial hula hoop' around the supernova site, has proposed that these faint outer rings might have been the result of two phenomena. First, there could be jets of material coming from an unseen neutron star or black hole very close to the supernova. These jets compressed part of the shell of gas around the blast site into a circular shape. Secondly, when radiation from the exploded star hit the rings, they lit up to form the ghostly images we see in images of the supernova. McCray disagrees with this interpretation, suggesting that Burrows' theory violates the 'tooth fairy rule' (which allows a credible theory to invoke only one mysterious, unknown and unseen agent to explain a rare phenomenon). However, both scientists agree that there is not yet a better explanation for these gracefully beautiful rings.

An independent indication of the expanding gas and the impending collision comes from the Space Telescope Imaging Spectrograph data shown in Figure 3.43b. The spectral image (middle panel) was taken through the aperture marked in the upper panel. The lower panel indicates the features in the spectral image. For Lyman-alpha, the hydrogen feature at left, the square was used to mask out radiation from the Earth's upper atmosphere. The broad feature

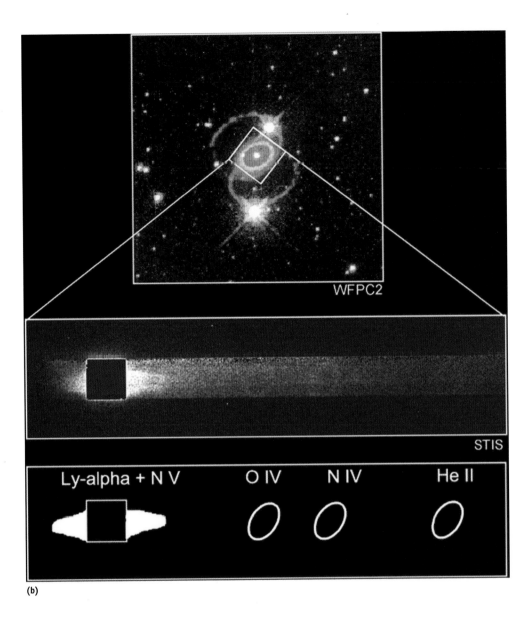

(b)

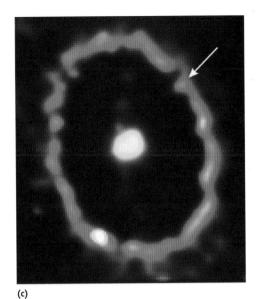

(c)

(d)

reveals gas traveling at 15 000 kilometers per second or 5 per cent of the speed of light. The expanding gas is slamming into gas lost by the star long ago. The collision between the expanding gas and the main, inner ring is estimated to occur (from updated data) around the year 2005. The spectral image also shows the inner ring in the light of the ions of oxygen (O), nitrogen (N), and helium (He).

According to current theories, there should be a pulsar showing up at the heart of SN 1987a, but observations by the High Speed Photometer showed no evidence for a pulsar, the rapidly spinning remains of a supernova's stellar core. As the core rotates, sometimes many thousands of times per second, a pulse of light flashes out like a beacon shining from a lighthouse. There should be a pulsar at SN 1987a, so why are astronomers not seeing it? There could be several reasons: (i) the pulsar is there, but obscured by the expanding cloud of debris; (ii) the remnant of the star may be too massive (i.e. greater than about three solar masses) to become a neutron star and has collapsed to a black hole; or (iii) the Earth may simply be out of the line of sight of the pulsar beam and we are just missing seeing it. Joe Dolan, a Goddard astronomer on the High Speed Photometry team, has speculated that the brightness for the leftover star may be very low indeed. He has estimated a visual magnitude of no more than 27 for the pulsed radiation from the supernova remnant. Any possible pulsar is certainly very faint, if it indeed exists.

Figure 3.43. (b) (top left) The spectral image of Supernova 1987a from STIS provides new information on the expanding gas around the dying star. See text for discussion. *(G. Sonneborn, Goddard Space Flight Center; J. Pun, National Optical Astronomy Observatory; NASA)*

Figure 3.43. (c–d) The top image **(c)** shows the ring as it appeared in 1994. **(d)** Shows the ring as it appeared in late 1997. The bright knot on the upper right side of the ring is the region where the blast wave, moving out from the site of the explosion, has collided with a circumstellar ring of material. The remains of the exploded star can be seen in the center. *(Peter Garnavich, Harvard-Smithsonian Center for Astrophysics; NASA)*

Figure 3.44. The Crab Nebula **(above)** as photographed from Palomar Mountain. The central
region is imaged by WF/PC-2 **(opposite)**, showing the pulsar (left member of the pair) and
dynamic knots and wisps. The changes are illustrated in Figures 3.45 and 3.46. *(Jeff Hester and
Paul Scowen, Arizona State University; NASA)*

Stellar death throes

The Crab Nebula Pulsar

The supernova that produced the Crab Nebula did form a pulsar – a rapidly spinning neutron star. The pulses are believed to be generated by electrons being whirled around in the strong magnetic field of the star. As these particles are hurled into the surrounding gases by the spinning of the magnetic field, they radiate energy, producing synchrotron radiation. This radiation is what causes the Crab to glow.

What appears to be a torus of gas surrounds the pulsar itself, and one of the x-ray jets is forming a ghostly looking halo of synchrotron radiation. In the direction of the opposite jet lies a shocked gas region where stellar winds collide with clouds of gas. Clearly, jets from the star are interacting with the clouds of gas and dust ejected when it exploded more than 900 years ago. University of Arizona scientists Paul Scowen and Jeff Hester have determined that the cooler regions of the nebula coincide with the thickest clouds of gas and dust. The hotter

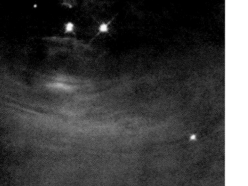 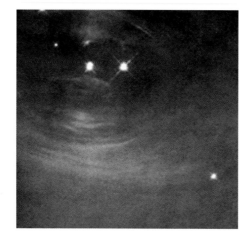 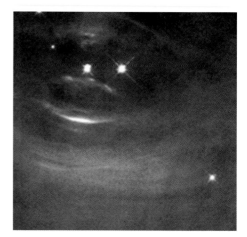

Figure 3.45. A time sequence of images showing changes in the Crab Nebula near the pulsar (left member of the pair) that powers the nebula. Features in the image are illustrated in Figure 3.46. *(Jeff Hester and Paul Scowen, Arizona State University; NASA)*

Figure 3.46. Schematic of the environment around the pulsar at the heart of the Crab Nebula. A vertical section is shown in the inset **(b)**, while the main drawing **(a)** is a perspective view. The 'halo' remains stationary but varies in brightness. The wisps move outward about one-half the speed of light. The jets are also pushing against and moving material, and the collision with this slower-moving material produces the highly variable 'sprite'. *(Jeff Hester and Paul Scowen, Arizona State University)*

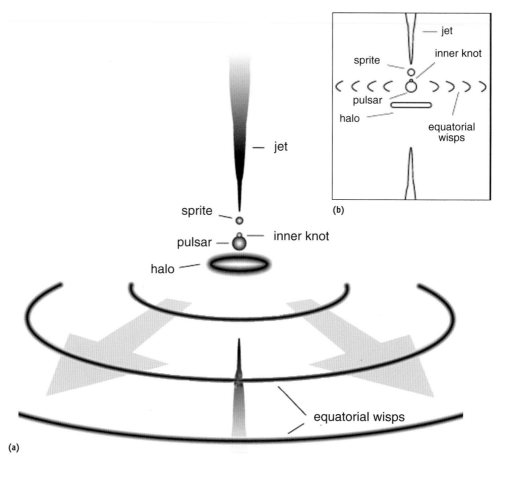

areas contain mostly thin, tenuous gas clouds. Several images taken at different times show that the region near the pulsar is undergoing dramatic changes.

The sensitivity (resulting in good temporal resolution) and wavelength coverage of HST's High Speed Photometer allowed a test of how the pulses of radiation travel through the surrounding shells. The HSP team wanted to see if the arrival time of the main pulse would coincide with the arrival times of all wavelengths of light – from gamma rays to infrared radiation.

The HSP obtained observations of the Crab Nebula pulsar in visible light (which is in the 4000 to 7000 angstrom range) and in ultraviolet light (in the 1700 to 2900 angstrom range). What HST 'saw' in the middle of the Crab Nebula was a pulsar ticking away 30 times a second, and both wavelength ranges showed the same arrival time. (Figure 3.47)

This is good news because it is a confirmation of the standard model for pulsed light emission. The positive check supports the position that we understand the production of pulsed radiation from rapidly rotating neutron stars, and, of course, the near-simultaneous arrival times for the light.

Supernova 1006

As we have seen with Supernova 1987a and the Crab Nebula, supernova remnants – the rapidly expanding shells of gas and dust formed when a dying star spreads its rich mixture of heavy elements out into the interstellar medium – are some of the most beautiful objects in astronomy. The remnant from the supernova of AD 1006 is an astronomical curiosity. A small-radius hot star is located behind the remnant, and the line of sight passes within about 2.5 arcminutes of the center of the remnant. This arrangement permits a study of the remnant through absorption lines produced when light from the background star passes through the remaining gas and dust.

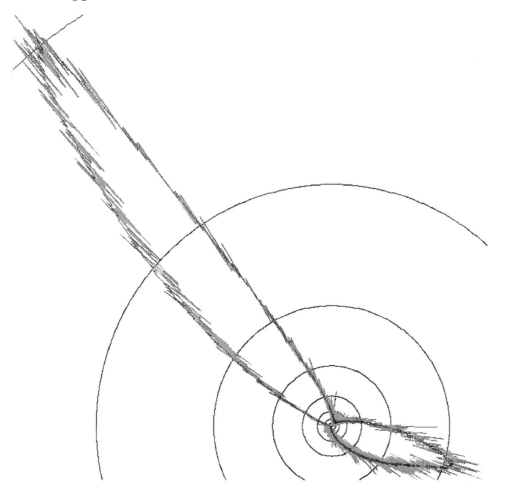

Figure 3.47. HSP measurements of the Crab Nebula pulsar plotted in a format where a complete revolution (360 degrees) is one period, and where the length is proportional to intensity. Visible light is plotted in red, and ultraviolet light is plotted in green. The curves in both wavelengths are remarkably similar. *(Robert Bless, University of Wisconsin and the HSP Team)*

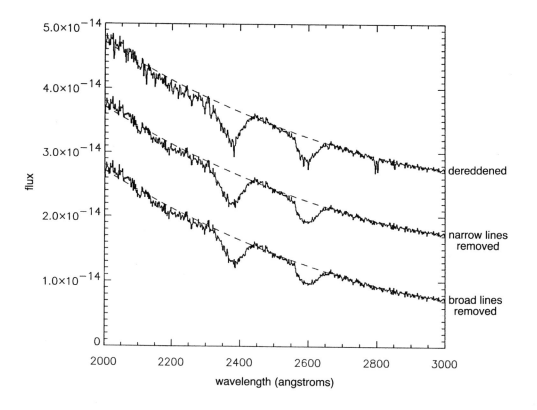

Figure 3.48. FOS spectrum of a hot star located behind the remnant of Supernova 1006. The different curves show the kinds of adjustments made to obtain the true profile of the iron (Fe II) lines. The bottom curve is the final result; the upper two are displaced for clarity. See the text for an interpretation of these absorption lines. *(Chi-Chao Wu, Computer Sciences Corporation)*

The Faint Object Spectrograph spectrum showing two broad absorption lines produced by once-ionized iron is given in Figure 3.48. From these lines we can make the following deductions about the material rushing out from the star: (i) the lines are symmetrical around their center, which implies that the original explosion was symmetrical – that is, the star exploded out equally in all directions, and (ii) the width of the line is produced by material receding away from us at about 8000 kilometers per second, and by material approaching us at the same speed. Using spectra to characterize the size and shape of this supernova explosion, as well as its speed and the chemical mix of the material rushing out from the star, has allowed astronomers to gain more insight into the complex dynamics of supernova explosions.

The Cygnus Loop

The last supernova remnant that we will study here is a beautiful arc of light called the Cygnus Loop. It stretches about 3 degrees across the sky in the constellation Cygnus, and lies in the plane of the Milky Way some 2600 light years away. The Loop is a blast wave from a supernova explosion estimated to have occurred about 15 000 years ago. Figure 3.49 shows the part of the loop known as the Veil Nebula.

The WF/PC-1 image in Figure 3.50 shows a small part of the Loop at better resolution, resolving details that are about the size of our Solar System. The blast wave from the supernova is slamming into clouds of interstellar gas. The shock process heats the gas and causes it to glow. Ejecta from the supernova envelope may be catching up with the blast wave, which has been slowed by interactions with the ambient gas. This gas is the bluish ribbon of light stretching across the image.

Eta Carinae, the cataclysmic variable star

The Milky Way Galaxy has a number of stars called cataclysmic variables, part of a larger class of variable or unstable stars, and we bring our exploration of stars to an end with one that has not quite died yet. This is the luminous blue star Eta Carinae, and one look at it would be enough to convince anyone that this star is in its death throes.

Eta Carinae is a Southern Hemisphere object, and has been quite familiar to observers throughout the centuries. Edmond Halley watched it brighten up to 4th magnitude in 1677. Its

Stars and the interstellar medium

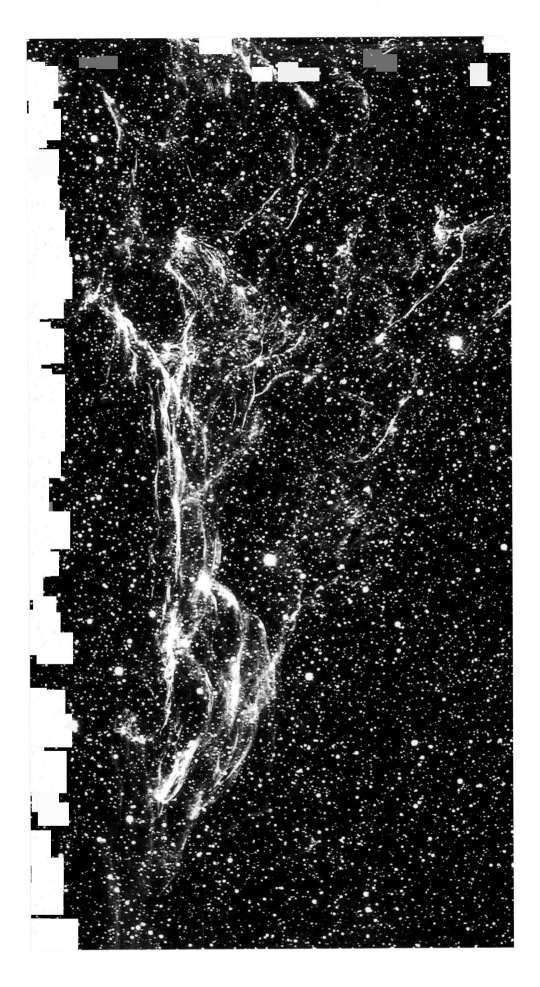

Figure 3.49. The Veil Nebula (NGC 6979) in the constellation Cygnus as photographed by the 4-meter Mayall Telescope of the Kitt Peak National Observatory. The nebula is also known as the Cygnus Loop. The image shows the north-central portion of the nebula. *(National Optical Astronomy Observatories)*

last outburst was in the 1840s, when it flared up to become the second brightest star in the sky. A.M. Clerk wrote in 1905 about the apparition:

> … *after a partial decline and several preliminary 'flutterings' it reached a final maximum in April 1843 when Sirius alone among the fixed stars slightly outshone it.*

What we see today is a rapidly expanding shell of material, radiating out in 'lobe-like' structures from the central star as shown in Figure 3.51. The reddish glow is actually light from fast-moving nitrogen and other gases ejected from the interior of the star at more than 3×10^6 kilometers per hour. The bright white material is very dusty and reflects starlight back to us. For scale, note that the two lobes cover an area about the size of our Solar System.

Astronomers are still trying to explain the dynamics of Eta Carinae's explosion. It certainly looks like a supernova, but so far it exhibits few other characteristics that would label it as such. The violence of the outburst of this cataclysmic star leads one to wonder what it will look like when it really does blow up!

Stars outside our galaxy

The study of stars *not* part of the Milky Way Galaxy is important for several reasons. We would like to convince ourselves that these stars are like the ones near us with the same physical processes governing their behavior. Often properties of individual stars and star clusters are used to estimate distances to other galaxies. We need to be sure that we know the brightnesses of these stars or clusters. Finally, we study stars and clusters in other galaxies to be alert to situations that may not be common in our galaxy or which may be obscured from our view. The capabilities of HST open up a wealth of opportunities for the study of stars outside our galaxy, and in the following we discuss a few examples.

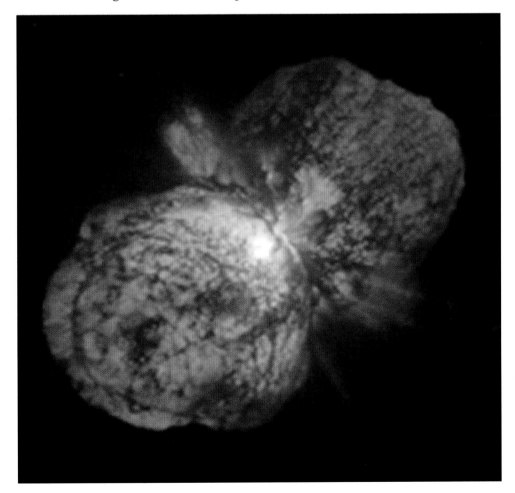

Figure 3.50. (Opposite) WF/PC-1 image of a part of the Cygnus Loop supernova remnant taken on April 24, 1991. The image is a color composite composed of blue light (oxygen atoms), green-light (hydrogen atoms), and red light (sulfur atoms); the different atoms arise from gas at different temperatures. The image shows remarkable detail in the nebula. *(Jeff Hester, Arizona State University; NASA)*

Figure 3.51. WF/PC-2 color composite image of Eta Carinae, a highly unstable star. Its last outburst occurred in the 19th century when, in 1843, it became the second brightest star in the sky. *(Jon Morse, University of Colorado; NASA)*

Figure 3.52. A color composite WF/PC-2 image of two young star clusters in the Large Magellanic Cloud. See text for discussion. *(R. Gilmozzi, STScI and ESA; S. Ewald, NASA-Jet Propulsion Laboratory; NASA)*

Figure 3.53. This WF/PC-2 image shows a giant starburst region, NGC 604, in the nearby galaxy M33. See text for discussion. *(Palomar Observatory; California Institute of Technology; Digital Sky Survey; H. Yang, University of Illinois; NASA)*

Stars in the Large Magellanic Cloud

Figure 3.52 shows an image of an area in the Large Magellanic Cloud (in the constellation Doradus). What appears to be a single cluster in the image turns out to be two distinct clusters at a distance of 166 000 light years. About 60 per cent of the stars belong to the dominant yellow cluster called NGC 1850, which has an age estimated at 50 million years. A scattering of white stars, about 20 per cent of the total, belong to another cluster estimated to be only 4 million years old. The rest are field stars in the Large Magellanic Cloud.

The situation of two well-defined star systems separated by only a small gap of space suggests that supernova explosions in the older cluster probably triggered the birth of the younger cluster. The Large Magellanic Cloud stars have a lower abundance of heavy elements than in our galaxy. Thus, their composition is more primordial, like the stars that formed early in the universe. A preliminary evaluation of the data indicates many more massive stars than expected. If this were also true of very early star formation, by some estimates it would have altered drastically the early history of the universe.

Stars in M33

Starbirth regions in other galaxies are expected to shed light on the intricacies of star formation across the universe. In one such place, the giant starbirth region NGC 604, in our neighboring galaxy M33, more than 200 hot stars have formed with masses in the range 15 to $60M_S$. These stars heat and illuminate the nebula like a lantern in a cave.

Starbirth nebulae are common in galaxies, and this is one of the larger ones, being nearly 1500 light years across. Because of its relative closeness to us (2.7 million light years away), we can have an excellent laboratory in which to observe the beginnings of stellar evolution.

Stars in M31

Figure 3.54 is an image of a globular cluster in the Andromeda galaxy (M31) known as G1 or Mayall II. This is the brightest globular cluster in M31 and contains at least 300 000 stars. This crisp image is comparable to ground-based views of similar clusters orbiting the Milky Way. On average, however, the Andromeda cluster is nearly 100 times farther away.

The details in the image allow study of the fainter helium-burning stars. Their temperatures and brightnesses show that the Andromeda cluster and the oldest clusters in our galaxy have approximately the same age. They were formed shortly after the Big Bang, and offer astronomers a glimpse of the earliest era of galaxy and star formation. Astronomers expect to study about 20 more clusters in Andromeda during the next few years.

Intergalactic stars

It is worth mentioning stars that exist outside of galaxies. These are intergalactic stars, stellar outcasts that were tossed out of their home galaxies by gravitational interaction into the dark emptiness of intergalactic space. The discovery of such objects was made by taking an exposure of a 'blank field' in the Virgo cluster of galaxies. The position was in the general vicinity of the elliptical galaxy M87 (Figure 4.17), but far enough away that any stars found would not be a part of M87's halo. The Virgo field was compared to the Hubble Deep Field

Figure 3.55. The vista from a hypothetical planet orbiting an intergalactic star. Only the spiral and elliptical shapes of galaxies (including M87 with its jet) appear in the sky because other individual stars are too far away to be resolved. *(J. Gitlin, STScI)*

Stars outside our galaxy

image (HDF; see Chapter 4), which was taken in a part of the sky devoid of any nearby cluster of galaxies. Comparison showed a major excess of stars, some 600 sources, in the Virgo image. These are bright, red giant stars. Harry Ferguson, from the Space Telescope Science Institute, has noted: 'These stars are truly intergalactic because they are so isolated that their motion is probably governed by the gravitational field of the cluster as a whole, rather than the pull of any one galaxy.'

Fainter stars should also be present in intergalactic space, but they are below the limit of Hubble's sensitivity. We are used to a night-time sky filled with stars, but this would not be true from a hypothetical planet orbiting an intergalactic star. Figure 3.55 shows the intergalactic star, but no other stars–they are too far away to be visible. The view contains only galaxies in the Virgo cluster. The bright elliptical galaxy is M87 and the visible jet of high-speed particles is shown (page 148, Chapter 4).

From the first brilliance of a newborn star to the last gasp of a dying supergiant, to the exile of stars from any known galaxy, HST follows the threads of stellar creation and destruction. The 'zoo' of stars we have visited in this chapter are just a very few of the more unusual stellar denizens to be found, and represent some of the most interesting examples of the lives of stars. Yet, understanding the differences and similarities between them – these variations on a stellar theme – prepares us for the wider task of studying the galaxies. HST observations, which are revealing the secrets of stars, are also starting to show astronomers the dynamics of large-scale stellar systems – and, as we shall see in the next chapter, the theme of creation and destruction is played out in grand scale across the face of the universe.

Galaxies

4

Images of galaxies are among the most beautiful and evocative pieces of cosmic artwork we can imagine. Open any coffee-table book of astronomical photography and you will marvel at the splendor of these collections of stars, the stellar cities of the universe.

In Chapter 3, we pointed out that studies of stars are limited necessarily by our own short-term perspective on them. We do not live for millions of years, and thus rarely see stars change drastically over time. The rare occasions when we do afford us the exquisite sight of an expanding supernova cloud or the eerie sight of a planetary nebula. It stands to reason that we do not see much change in galaxies with their multi-*billion* year lifetimes. What we see of galaxies are – at best – snapshots of stellar conglomerations, frozen in time. The time we see is in the past, for the farther out from Earth we look, the further back in time we are seeing.

The realization that stars are organized into galaxies is a comparatively recent one, and it parallels the gradual discovery of our own place in the cosmos. Of course, the starry swath that the Milky Way traces out across the sky has evoked awe and wonder in the human mind since prehistoric times, but it was not until the early seventeenth century, when Galileo turned his telescope to it, that astronomers saw that the Milky Way was not a smooth pathway of light but a myriad of stars. The idea that we were looking edgewise through 'our own galaxy' was – and perhaps still is – completely astounding to some.

Even the words used to describe galaxies through the years illustrate the confusion about just what they really were. The term 'nebula' was applied to any fuzzy cloud-like structure in the night-time sky that obviously was not a star. Compounding the problem was that some fuzzy things in the sky actually moved over short periods of time. As it turns out, these were comets. (In fact, Charles Messier compiled his famous Messier catalog of 'fuzzy-looking things'

In a sense the galaxy hardest for us to see is our own… we are far out from the center, and to make matters worse, we lie in a spiral arm clogged with dust. In other words, we are on a low roof on the outskirts of the city on a foggy day.

Isaac Asimov

Immense as the Milky Way appears, it does not travel the universe alone. Other distant islands of light wheel through the cosmos. We find no two exactly alike— galactic snowflakes, drifting through the universe.

Carolyn Collins Petersen

Figure 4.1. The Andromeda Galaxy, M31, in the constellation Andromeda. This Sb spiral galaxy is the nearest and most easily visible spiral galaxy. This is a computer enhanced image of a plate taken at the Kitt Peak National Observatory. *(National Optical Astronomy Observatories)*

– which is where the 'M' numbers you see applied to some astronomical objects originate – so that comet seekers would not be confused by the permanent nebulae in the sky.)

Everywhere you look there are galaxies. Some are quite distant and hard to see without at least a fairly good-sized telescope, but others are easier to find. They come in a veritable rogue's gallery of shapes, sizes, and masses. And they evolve.

In the northern hemisphere sky, stargazers can find the Andromeda Galaxy, also known as M31, not far from the W-shaped constellation Cassiopeia. Figure 4.1 shows a ground-based image of the galaxy, which lies 2.2 billion light years from us and has more than 300 billion stars. In the southern hemisphere, stargazers delight in the vision of the Large and Small Magellanic clouds – companion galaxies to our own Milky Way.

Spiral galaxies are typified by M31, and by galaxies like the one seen in Figure 4.2. We see this galaxy in Coma Berenices 'edge-on', but we know it is like the Milky Way in many impor-

Figure 4.2. NGC 4565, an Sb spiral galaxy in the constellation Coma Berenices seen edge-on. *(D. DiCicco, Sky & Telescope Magazine)*

Figure 4.3. M81 (NGC 3031), an Sb spiral galaxy in the constellation Ursa Major showing prominent spiral arms. *(National Optical Astronomy Observatories)*

tant respects. Spiral galaxies rotate, taking hundreds of millions of years to make a complete revolution. We do not see this stately motion visually, but if we take spectra we can measure a redshift on the side of the galaxy moving away from us, while the side of the galaxy moving toward us will be blueshifted.

The arms of spiral galaxies are of special interest to us for a couple of reasons: (i) because our Solar System lies in a spiral arm of the Milky Way; and (ii) because spiral arms are prime sites for the birth of stars (for example, the stellar nurseries in the Orion Nebula). Figure 4.3 shows the galaxy M81, its spiral arms shining with the light of young and middle-aged stars. As a general rule, galaxies with more open arms tend to have more gas and dust, and show a greater tendency for brisk episodes of star formation. The central bulges of spiral galaxies assume different sizes and shapes, and generally have older type stars. Figure 4.4 shows galaxy NGC 4594; its central bulge is among the largest known.

Figure 4.4. M104 (NGC 4594), an Sa spiral galaxy with a large central bulge. The dark band across the galaxy's central region is composed of dust and gas. This is a 4-meter Mayall Telescope photograph from the Kitt Peak National Observatory. *(National Optical Astronomy Observatories)*

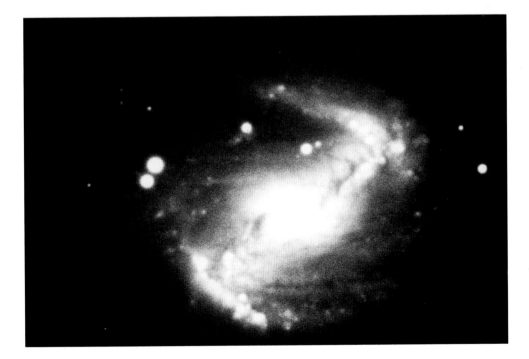

Figure 4.5. NGC 5383, a type Sbb, barred spiral galaxy in the constellation Canes Venatici. This type of galaxy has a bar of stars passing through the center. This is a 4-meter Mayall Telescope photograph from the Kitt Peak National Observatory. *(National Optical Astronomy Observatories)*

Figure 4.6. A view of the Large and Small Magellanic Clouds. *(National Optical Astronomy Observatories)*

Figure 4.7. The irregular and active galaxy M82. The composite image approximates the true appearance of the galaxy, and was taken by George Jacoby at the Kitt Peak National Observatory. *(National Optical Astronomy Observatories)*

Barred spiral galaxies are a special case. They have long bars of stars passing through their central regions; for an example, see Figure 4.5, which shows the galaxy NGC 5383 in the constellation Canes Venatici.

The Large and Small Magellanic Clouds (shown in Figure 4.6) are companion galaxies to the Milky Way. They are a pair of *irregular* galaxies that lie in the constellations Doradus and Tucana, respectively. M82 in Ursa Major (Figure 4.7) is another irregular galaxy. Think of them as chaotic systems of stars and interstellar material, with very little apparent structure. Irregulars undergo sporadic waves of starbirth, which can be triggered by supernovae and other events.

Lastly, we have the *elliptical* galaxies, named for their ellipsoidal appearance (see Figure 4.8). An elliptical galaxy shows a smooth outline, and its shape depends on is structure and orientation in space. Ellipticals are generally not very active galaxies in that they boast no regions of star birth, and indeed have very few young stars, or little of the gas and dust needed for star formation.

The evolution of galaxies

Because stars evolve, we strongly suspect that galaxies must also evolve. The problem is that any evolutionary sequences would take millions and millions of years – far too long for any type of meaningful real-time observations. However, there are ways around this problem. Clusters of galaxies can serve as the beginning of insight into evolution of galaxies. For example, we can study images of galaxies in clusters and see if the different types of galaxies are present and in what proportion. If the results do not match the nearby galaxies, evolution is the likely suspect.

Alan Dressler, of the Observatories of the Carnegie Institute of Washington, and a group of colleagues used both Wide Field and Planetary Cameras to make observations of a remote cluster of galaxies, CL 0939+4713. According to Dressler's data, the age of this cluster is about two-thirds of the universe's present age, or, put another way, about one-third of the way back to the Big Bang. Depending on how old the universe is, this cluster could be around 3–4 billion years old.

It seems that there are many more spiral galaxies in early clusters than are found in more recent clusters. This confirms previous studies, which implied that star-forming galaxies were more prevalent in the early universe. Because we know that spiral galaxies are prime sites of star formation, by studying them we can look back to a time when clusters of young galaxies

Figure 4.8. An E4 elliptical galaxy, NGC 4435, in the constellation Virgo. A peculiar spiral galaxy is also shown. *(National Optical Astronomy Observatories)*

Figure 4.9. Hubble's famous 'tuning fork' diagram showing his classification scheme for galaxies. *(Yerkes Observatory, University of Chicago)*

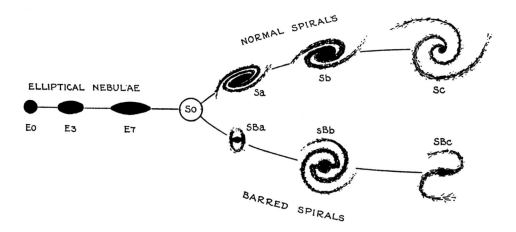

went through massive amounts of starburst activity – the formation of new stars. Galactic starbursts take place over large scales and often involve collisions of galaxies.

It also seems now that the number of interacting galaxies is large, demonstrating that the processes of disruption and merger are very important in the normal evolution of galaxies, as are the changes caused by stellar evolution.

These points are more significant if you look at a galaxy classification scheme proposed by Edwin Hubble in the 1920s: the famous 'tuning fork' diagram of galaxy shapes shown in Figure 4.9. The ellipticals are labeled according to their apparent ellipticity. The normal spirals vary from types labeled Sa, for those with strong central bulges and tightly wound arms, to types called Sc, for those with little central bulge and open arms. A similar sequence exists for the barred spiral galaxies. The class S0 is a hypothetical transition type. Some of these types were referred to as 'early' or 'late', designations with evolutionary connotations. Today we think the tuning fork diagram does not represent an evolutionary sequence. Hubble included other galaxies, such as Irr (for irregular), and the scheme has been updated from time to time; for example, astronomers have added a type Sd.

The galaxies in CL 0939+4713 have been arranged according to the scheme described above (see Figure 4.10), and from that work comes a startling insight into the nature of galactic evolution. Clusters of spiral galaxies, which were more numerous in the early universe, seem to undergo great changes as they mature. If we compare the galaxies in this distant cluster to what see in more contemporary clusters, we can see these changes.

The top three rows show elliptical galaxies, with various ellipticities, and possibly some S0 galaxies. These types of galaxies are common in nearby clusters at the present time.

Rows 4–7 show spiral galaxies with differing degrees of openness of the spiral patterns from Sa to Sd. The Sds in row 7 are shaped atypically – that is, they do not look like normal spirals. These are common in CL 0939+4713, but they are not common in nearby clusters today.

Row 8 provides another clue as to why the spirals are less common today. These are galaxies apparently merging into single galaxies. The interaction between galaxies – causing merging and disruption – must be considered along with a phenomenon called fading due to the effects of stellar evolution (stars are brighter when they are younger and dimmer as they grow old) when explaining why star-forming galaxies – the spirals – were more prevalent in young clusters than in the older contemporary clusters.

Fig 4.10. (Opposite) A HST classification of galaxies from the cluster CL 0939+4713, (Figure 1.15, page 19). The original image was obtained with WF/PC-1. See the text for a discussion. *(Alan Dressler, Carnegie Institution; NASA Co-Investigators: Augustus Oemler, Yale University, James E. Gunn, Princeton University, and Harvey Butcher, Netherlands Foundation for Research in Astronomy)*

The Hubble Deep Field (HDF): the survey

One of Hubble Space Telescope's most significant – and aesthetically beautiful – contributions to the study of galaxy evolution is a project called the Hubble Deep Field. It was first proposed by Institute Director Robert Williams and a team of researchers as a way to 'take a census' of typical galaxies in a single patch of sky.

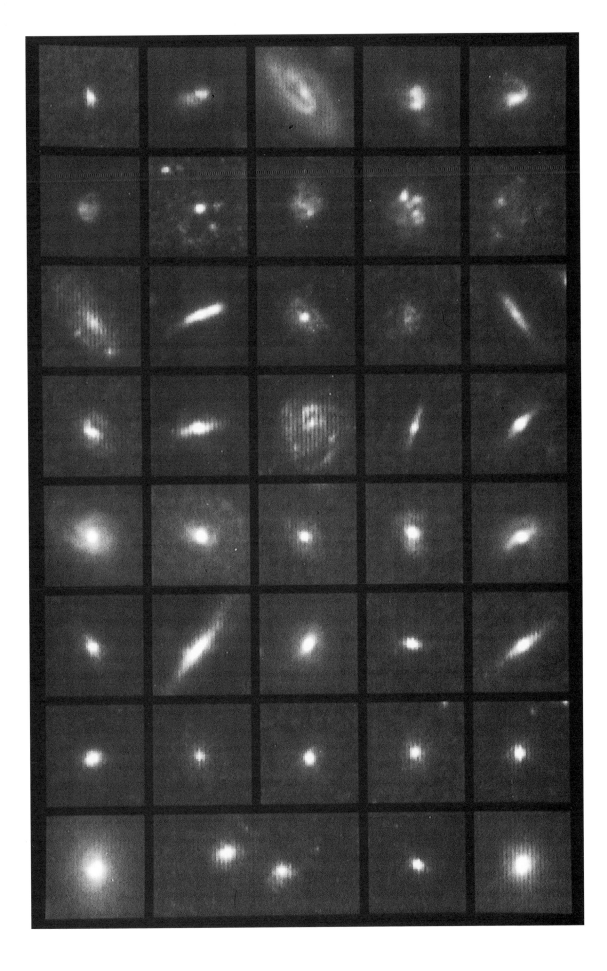

Figure 4.11. (Opposite) The full Hubble Deep Field was produced by observing continuously for 10 days. Three different filters were used to obtain this natural color image. This represents the longest HST exposure yet taken. *(R. Williams, STScI, and the Hubble Deep Field Team; NASA)*

Figure 4.12. (Above and left) Details of three regions in the Hubble Deep Field. The areas show a bewildering variety of galaxy shapes and colors. *(R. Williams, STScI, Hubble Deep Field Team; NASA)*

In December 1995, HST's WF/PC-2 was trained on a small region of the sky near the Big Dipper for 10 days. Hundreds of exposures were taken through filters centered roughly on 3000 angstroms (ultraviolet), 4500 angstroms (blue), 6000 angstroms (orange), and 8000 angstroms (far red). The particular region of the sky in Ursa Major was chosen for the relative absence of nearby stars, and for the fact that the telescope could observe the target area continuously throughout its orbit. The field contains nearly 3000 galaxies down to the faint limit of the images. The faintest galaxies in astronomical notation are visual magnitude 30, the equivalent of one photon per week hitting the human eye. The galaxies at the limit are about 15 times fainter than can be routinely achieved from the ground. Of course, the study of lensed galaxies will enable the study of fainter galaxies both from the ground and from HST.

The impetus for the HDF was a tantalizing situation in early 1995. Observing programs on clusters of galaxies and on other distant galaxies pointed toward important new understandings in the evolution of galaxies and cosmology. It was clear, however, that the surface had been just barely scratched. Robert Williams decided to devote a major fraction of his Director's Discretionary Time to a major effort on a single patch of sky, the HDF. A critical point of the campaign was the immediate dissemination of both the raw and reduced data to the scientific community. Analysis of the HDF images continues to be a major research activity by many groups, and plans have been made to do a southern hemisphere version of the deep field, the SHDF, in late 1998.

The authors first saw a presentation of the HDF at the January 1996 meeting of the American Astronomical Society in San Antonio, Texas, in the form of an 8-foot-high rendering of the

Figure 4.13. (a) Images in the far red, orange and blue, left to right, show a distant galaxy, but the ultraviolet image (far right) does not show the galaxy. It has 'dropped out'. *(M. Dickinson, STScI).* **(b)** The images, right to left, show no galaxy at the position marked by the arrow, in ultraviolet, blue and orange light, respectively; it does appear in the far red light image (far left). See text for estimated z values that produce these effects. *(K. Lanzetta, State University of New York; Stony Brook, NASA)*

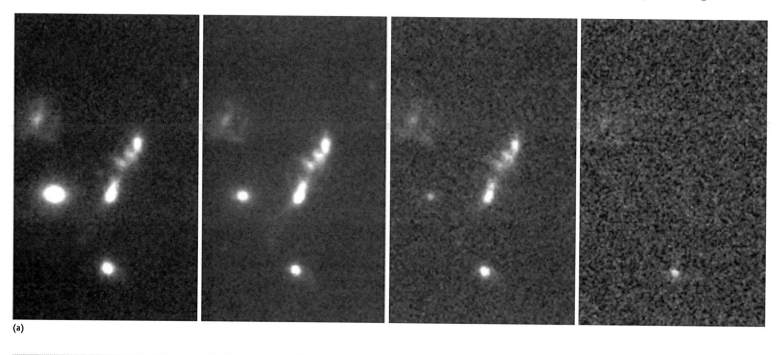

(a)

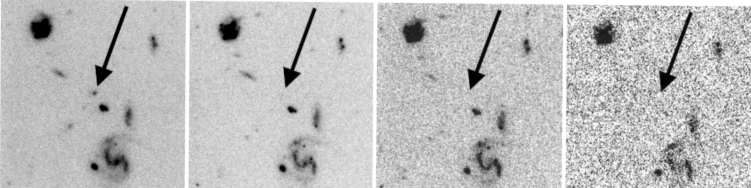

(b)

entire image. Many of us stood quietly before the image and let the beauty and profound nature of the view fill our field of view. The area covered by the image is a field 2.7 arcminutes on a side and covers an area 1/140 the apparent size of the full moon or 1/27 000 000 of the entire sky. The beauty of the view belies the incredible amount of data it contains.

Results relating to cosmology are discussed in Chapter 5, and those relating to the evolution of galaxies are discussed later in this chapter. Before we dive into that discussion, let us examine the observational technique for estimating the distances to the faint galaxies in the HDF.

Ultraviolet dropout

The images in Figure 4.13 clearly show that the visibility of distant galaxies *through different color filters* depends critically on the distance of the galaxy. Distances can be determined from the redshift of the spectrum via the Hubble constant (see discussion in Chapter 5). The larger the redshift of the spectrum, the larger the distance. Astronomers often quote the 'distance' of galaxies in terms of the quantity z, a measured quantity in the spectrum. The value of z gives the shift in units of the original wavelength. If a line was originally at (say) 2000 angstroms, $z = 3$ means a shift of 6000 angstroms or a new wavelength of 8000 angstroms; $z = 2$ produces a shift of 4000 angstroms or a new wavelength of 6000 angstroms; $z = 1$ produces a new wavelength of 4000 angstroms; while $z = 0$ is no shift at all.

The top images in Figure 4.13 show *no* galaxy visible in the ultraviolet image, but the galaxy is easily visible in the blue, orange, and far red filters. This means that this is a 'relatively nearby' distant galaxy with a z value of roughly 3. The bottom images show a galaxy visible *only* in the far red filter. This is a 'relatively distant' galaxy with a z value of roughly 6.

The method of determining the z value works as follows. The hydrogen atoms between the galaxies, called intergalactic hydrogen, can absorb light from galaxies. This hydrogen (along with everything else in the universe) is moving away from us and hence its absorption is redshifted too. The intergalactic hydrogen is a natural filter that cuts out the light from galaxies at longer and longer wavelengths as z increases. The specific z values are assigned from models of galaxy light that are calculated for different values of z and their comparison with the observations.

The technique may not be accurate for an individual galaxy, but should give a reliable overall view of the distances of galaxies in the Hubble images. The technique has already been checked through use of spectroscopic redshifts taken with the Keck telescope in Hawaii.

Rate of star formation

The knowledge of distances to the galaxies in the HDF permits us to make an estimate of the rate of star formation in distant galaxies. Hot, massive stars are an excellent tracer of starbirth activity and are all very bright in ultraviolet light. Not all of the original ultraviolet light makes it to our detectors: the light shortward of 912 angstroms is absorbed by intergalactic hydrogen, but ultraviolet light longward of 912 angstroms passes through. For distant galaxies, this break is redshifted into the visual wavelength region with the exact location depending on the z value (redshift) of the galaxy. Of course, this is the basic technique used to determine the z value as discussed when describing the ultraviolet dropout phenomenon. With the redshift known, the wavelength range for the redshifted ultraviolet light can be calculated. The brightness of the light in this range is a measurement of the relative rate of star formation.

In all cases, the point is to determine the redshift or z value from a change in the spectrum or the color, and to measure the brightness at a specific region in the spectrum that indicates the rate of star formation.

The HST results around cosmic time of 1 to 2 billion years after the Big Bang come directly from the HDF and were produced by a team led by Piero Madau at the Space Telescope Science Institute. Their work could not be extended toward cosmic age 5 billion years with the technique described above because the redshifts were not large enough to shift the break in the

Figure 4.14. The star-formation rate as a function of redshift, z. The data are from Gallegos et al. (1995), red triangle; Lilly et al. (1996), dark blue circles; Connolly et al. (1997) light blue squares; Madau et al. (1996) and Madau (1997), red squares. The agreement between the different studies is very encouraging. The basic result is that star formation today ($z = 0$) is much less than in the past. The peak was at $z \sim 1.5$ (or 5 billion years cosmic time or time after the Big Bang) at about 12 times the current rate. Remember that z is the observed quantity. Conversion to cosmic time was done by using the Einstein–deSitter relationship, $t = T_0/(1 + z)^{3/2}$, with $T_0 = 13$ billion years. (P. Madau, STScI)

Figure 4.15. (Opposite) (a) These are traditional spiral and elliptical galaxies that make up the two basic classes that we see in our current epoch, about 8–13 billion years after the Big Bang. Elliptical galaxies contain older stars, while spirals have vigorous on-going star formation in their disks. Our Milky Way is a typical spiral. Both galaxies in this column are a few tens of millions of light years away. **(b)** These galaxies existed in a rich cluster when the universe was approximately two-thirds its present age. Ellipticals (top) appear fully evolved because they resemble today's descendants. By contrast, some spirals have a 'frothier' appearance, with loosely shaped arms of young star formation. The spiral population appears more disrupted due to a variety of possible dynamical effects from dwelling in a dense cluster. **(c)** Distinctive spiral structure appears more vague and disrupted in galaxies that existed when the universe was one-third its present age. These objects do not have the symmetry of current-day spirals, and contain clumps of starburst activity. However, even this far back in time, the elliptical galaxy (top) is still clearly recognizable, but the distinction between ellipticals and spirals grows less certain with increasing distance. **(d)** These extremely remote objects existed when the universe was one-tenth its present age. The distinction between spiral and elliptical galaxies may well disappear at this early epoch. However, the object in the top frame has the light profile of a mature elliptical galaxy. This implies that ellipticals formed early in the universe, whereas spirals took much longer to form. (A. Dressler, Carnegie Institution of Washington; STScI; D. Macchetto, ESA and STScI; M. Dickinson and M. Giavalisco, STScI; NASA)

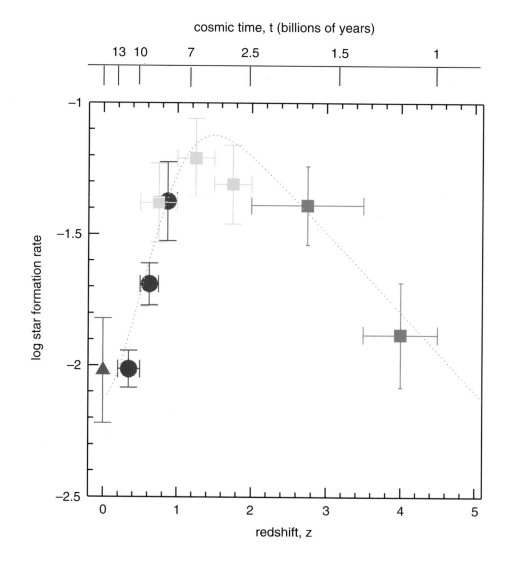

spectrum into the visual wavelength region. The next range was filled by an approach used by Andrew Connolly and Alexander Szalay, Johns Hopkins University, and Mark Dickinson, Space Telescope Science Institute. They examined galaxies in the Hubble Deep Field with the 4-meter telescope at the Kitt Peak National Observatory. The redshift was determined from the galaxy's infrared light, light not picked up in the HDF exposures. The results from cosmic time of 5 billion years to near the present day were obtained from the extensive ground-based Canada–France Redshift Survey (CFRS) led by astronomer Simon Lilly, University of Toronto, Canada. The limit of this survey is the faintness of the galaxies earlier than cosmic time of 5 billion years (or 8 billion years lookback time for an adopted age of 13 billion years for the universe). Finally, the current rate of star formation ($z = 0$) was determined by a team led by J. Gallego.

The results from this analysis of the Hubble Deep Field are that: (1) the rate of star formation today is much less than during most of the history of the universe; and (2) the peak rate was about 12 times the present rate when the universe was approximately one-third its present age, or at a cosmic time between 4 and 5 billion years.

The theory of star formation in galaxies now needs to be tested against the new results. Among the questions to be answered are: What types of galaxies are responsible for the peak in star formation? What are the locations of star formation in these galaxies? Does our knowledge of stellar evolution apply in detail to stars in early galaxies? Answers to some of these questions should be more easily determined by using the Space Telescope Imaging Spectrograph, which can study galaxies five times fainter than those seen in the Hubble Deep Field data.

Evolution of galactic forms

The rate of stellar births discussed in the previous section provides a target for theorists to explain how the stars in galaxies are formed and evolve. But, do the forms of galaxies change, and, if so, are the changes important in the grand scheme of things? When we discussed the Hubble classification scheme for the cluster CL 0939+4713, we saw strong evidence for changes in galactic forms.

As HST directs its gaze toward more of these distant galaxy clusters, what it finds offers tantalizing clues to the evolution of galaxies. Figure 4.15 shows snapshots of galaxies ranging in age from current time back to an epoch when the universe was a small fraction of its current age, and makes an excellent illustrative summary of galactic evolution:

(1) the ordinary changes responsible for the formation of galaxies by gravitational attraction and flattening due to rotational processes;
(2) the explosions responsible for activity in galaxies;
(3) the normal formation and evolution of stars. Massive stars will evolve quickly and recycle part of their masses into the interstellar medium in the form of gas and dust, and new stars will form in these regions. Even if the mass stays in the same general area, the appearance must change;
(4) disruption and merging through collisions, which should drastically alter their structure and appearance.

Piecing the various possible processes together to produce a generally accepted, detailed picture of the evolution of galaxies is a major challenge. Any theory will be tested by comparison with galaxies at different distances, which, because of light-travel time, is equivalent to looking back at galaxies in earlier and earlier times.

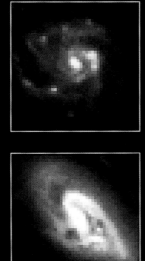

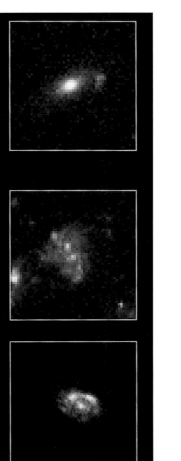
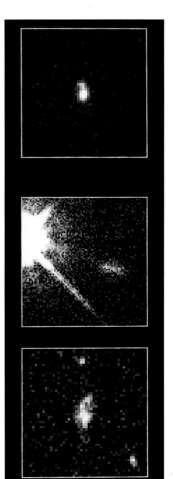

(a) (b) (c) (d)

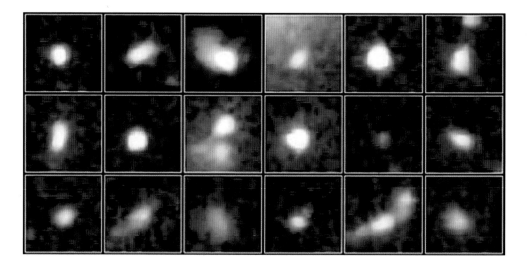

Figure 4.16. True-color images of 18 possible galaxy building blocks from WF/PC-2. Their distance is approximately 11 billion light years, and they shine from the light of several billion stars. The sub-galactic clumps could have merged to form the galaxies we observe today. *(R. Windhorst, Arizona State University; NASA)*

Before galaxies

How do the galaxies we observe today as fascinating giant and luminous structures come about? They were probably not formed as a uniform whole (with approximately their present mass), but began as sub-galactic clumps that collided and merged to form galaxies.

This view receives support from observations of a patch of sky in the constellation Hercules, near the border of Draco. Exposures were taken in blue, green, and far-red light during 48 orbits of HST around the Earth. These exposures were combined to produce the true-color images shown in Figure 4.16. Eighteen of the objects are shown . They shine from the ultraviolet light of several billion stars that is redshifted into the visible light range. Ten of the sub-galactic clumps have ground-based spectra which confirm their distance of 11 billion light years. Each is some 2000 to 3000 light years across. Additional support for the building block role is the fact that these clumps are more numerous in a given volume of space than today's more luminous galaxies.

Thus, we see the beginnings of galaxy formation with the clumps that merged over time to form the sizes and masses now, typically found. The density in the earlier universe was much higher than now and so collisions were more frequent. As the galaxies formed and evolved, they created the changes in star-formation rate and the shapes of galaxies that we have just described. Now the challenge is to confirm the picture and provide quantitative understanding.

Individual galaxies

It is hard to believe when we look at a galaxy in the sky, or an image in a book, that some galaxies are host to some of the most frenetic activity in the universe. For a long time, galaxies were thought to be quiescent. That changed in 1943, when astronomer Carl Seyfert drew attention to a handful of galaxies with unusual properties. Now called Seyfert galaxies, they have strong emission lines from their central regions and a starlike appearance. Ordinary galaxies have only absorption lines and inconspicuous, sometimes 'blobby', nuclei that more or less blend in with the rest of the central region. It was a good bet that something energetic and unusual was going on in Seyfert galaxies.

Astronomers now recognize huge variations in the activity levels of galaxies. There is the level we observe in our own galaxy: strong radio signals emanating from the nuclear region of the Milky Way. A middle level shows clear evidence of explosive activity from nuclear regions of other galaxies; the Seyfert galaxies are the most active example of this type of galaxy. The highest level of activity includes the superluminous quasi-stellar objects – quasars, for short. These are probably the most exciting candidates for galactic studies with HST.

Galaxies with central black holes

Not long ago, black holes were an interesting theoretical construct, despite the tendencies of Hollywood sci-fi movie makers and science-fiction writers to use them as integral parts of stories involving space travel and intergalactic intrigue.

Speculation about an object so dense that light could not escape from it goes back at least two centuries, but a formal understanding required the Theory of General Relativity as developed by Albert Einstein at the beginning of the 20th century. Today, astronomers and physicists are convinced that black holes exist, and they use them to explain many rare and exotic happenings in quasars and galactic nuclei. Black holes seem to be an important constituent of the universe, and a short review of their fascinating properties may be useful before we look at what HST has seen in its search for these exotic stellar beasts.

Although the matter in a black hole has collapsed to a point, the effective 'size', or boundary, of the object is defined by the event horizon. Because light cannot escape from a black hole, happenings within the event horizon cannot be known to an observer on the outside, who can only speculate based on what is occurring outside the black hole. The size of the event horizon itself is calculated by taking the mass of the black hole (in solar masses) and multiplying it by 3 kilometers.

According to theory, black holes have only three properties: their mass, their spin (a measure of rotation), and their electrical charge. The electrical charge is believed to be zero, and so the properties are even simpler. For our purposes, the property of interest is the mass, and we find that black holes come in three varieties: mini, stellar, and supermassive.

Mini black holes could have formed in the immense pressure, temperature, and turbulence that existed shortly after the Big Bang. They would have masses comparable to mountains and, after billions of years, would vanish in a flash of energy. The properties of these mini black holes are governed by general relativity and by quantum mechanics. This accounts for their atypical behavior, which is to emit light, something the larger black holes do not do. While mini black holes may be quite common in the universe, there is currently little direct evidence for their existence.

Stellar black holes are formed by massive stars at the end of their lives. When a star with a mass equal to $3M_S$ or greater exhausts its internal energy source, no known force can stop the star's collapse, and it forms a black hole. There is considerable evidence for the existence of stellar black holes.

Supermassive black holes are probably an essential component of the cores of many galaxies. Conditions at the centers of galaxies seem ideal for the formation of massive stellar black holes, which have merged or gathered up material to form larger and larger black holes. Supermassive black holes can grow to millions or billions of solar masses, and there is good evidence that they exist. In addition, supermassive black holes are the most likely viable source of energy that can explain the superluminosity of quasars.

You may be asking yourself the following question: if no light escapes from black holes (except for the minis), how can we see them? The answer is that the gravitational field near the event horizon is very strong, and material falling into the black hole gains energy and is compressed. Thus, we 'see' black holes by observing the hot, dense material around them; this material is a strong emitter at ultraviolet, x-ray, and gamma-ray wavelengths. Evidence for the existence of a specific black hole is usually circumstantial, and not every emitter of high-energy radiation is necessarily a black hole. Generally, astronomers are comfortable when they see numerous effects that would require the existence of black holes to explain them.

Many galaxies have some form of enhanced emission – radio, infrared, ultraviolet, or x-ray – and often jet-like features are observed. The possibility that the 'central engine' in the form of a massive black hole powers the activity is a good one, and imaging results on the central regions of several galaxies support this. Based on a growing body of HST data, it turns out that central massive black holes may be rather common in galaxies.

Figure 4.17. (Below and opposite) The gas disk in the nucleus and jet of M87, as imaged by WF/PC-2. *(H. Ford, STScI and Johns Hopkins University; Z. Tsvetanov, A. Davidsen and G. Driss, Johns Hopkins University; R. Bohlin and G. Hartig, STScI; R. Harms, L. Dressel and A.K. Kochhar, Applied Research Corp; B. Margon, University of Washington-Seattle; NASA)*

The matter around black holes seems to form two common features – disks and jets. When material contracts over astronomical distances, it forms rapidly rotating disks; we expect to see these around black holes. The disks can be quite thick and dense in the plane of the disk, but the central part, perpendicular to the disk, can be relatively transparent. Thus, energetic phenomena produced by the material falling into the black hole cannot escape through the disk, but they can escape perpendicular to the disk, thus permitting the formation of 'jets'. Recall the discussion in Chapter 3 for the contraction of gas and dust clouds to form stars. Disks and jets seem to be a common feature in the universe. This is exactly the situation that HST often finds in the centers of galaxies.

M87

Within the Virgo Cluster of galaxies lies one of the first known 'peculiar' galaxies – a giant elliptical galaxy identified as M87. It has been known for years to have a bright, optical jet in its central region. In 1991, the Faint Object Camera took an image that reveals unprecedented details. The jet was found to be some 40 000 light years long, and it has features as small as 10 light years across. At the time, astronomers suspected that the jet was emanating from a massive black hole. The WF/PC-2 view of the jet and core of M87 is shown in Figure 4.17. The faint starlike sources around the central region are globular clusters, each containing between 100 000 and 1 000 000 stars.

In 1994, HST provided what many astronomers consider conclusive evidence for a black hole at the heart of M87. The evidence is based on velocity measurements of a disk of hot gas whirling around the black hole. The presence of the disk in the images allowed for precise measurement of the black hole's mass, which was discovered to be as much as 3 billion Suns ($3 \times 10^9 M_S$), but is concentrated into a space the size of the Solar System. The speeds of the, mostly hydrogen, gas in orbit around the object are *tremendous* – at least 550 kilometers per second! This indicates that something very massive is in the center, and since there are few stars there, it must be a black hole. The Faint Object Spectrograph studied the spectral signature of the orbiting gas, and that observation was the clinching evidence for the black hole (see Figure 1.18, page 23).

Holland Ford, of the Space Telescope Science Institute and Johns Hopkins University, and Richard Harms (who was at Applied Research Corporation at the time) were co-investigators on the observation. Harms actually thinks that the center of M87 could harbor something much stranger than a black hole, but it is hard to imagine what that might be. 'A massive black hole is actually the conservative explanation for what we see in M87,' he said. 'If it's not a black hole, then it must be something even harder to understand with our present theories of astrophysics.'

NGC 4261 and M84

Figure 4.18 shows an image of the central region of the galaxy NGC 4261 The existence and appearance of a dusty disk strongly implies a black hole at its center. Doppler shift measurements of the disk (which has a mass of $10^9 M_S$) show that it is rotating at 420 kilometers per second, strong evidence for the existence of yet another supermassive black hole. The article reviewing this work in *Science* carried the headline, 'Just Another Billion-Sun Black Hole!' – a humorous attempt to illustrate how often black holes are being found these days. Similar results for galaxy NGC 3115, from data taken by a team of astronomers led by John Kormendy, University of Hawaii, imply a central black hole for that galaxy with a mass of 2 billion Suns!

Figure 4.18. WF/PC-2 image of the central region of NGC 4261. This image shows fine structure in the disk, which may be produced by waves or instabilities. *(H. Ford and L. Farrarese, Johns Hopkins University; NASA)*

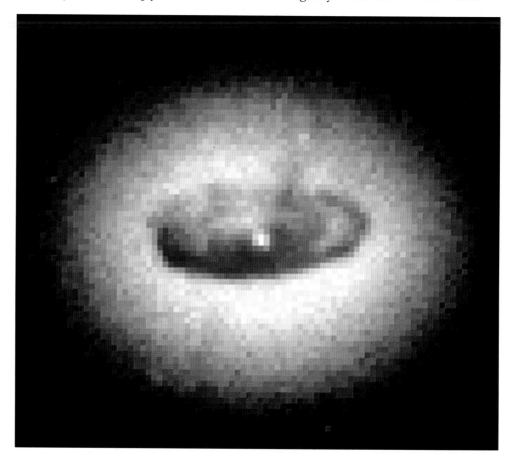

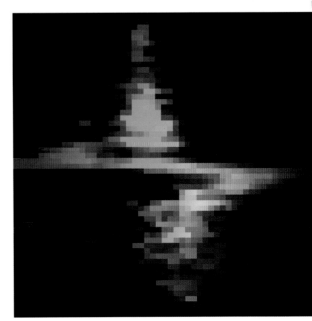

Figure 4.19. The STIS spectral image **(above)** of the region (blue box) in the WF/PC-2 image **(left)** of the nucleus of galaxy M84 carries the signature of a central black hole. The zigzag color pattern is explained in the text, but shows, at a glance, the existence of the black hole. *(G. Bower, National Optical Astronomy Observatories; NASA)*

With the installation of the Space Telescope Imaging Spectograph (STIS) on HST, the search for central supermassive black holes can be carried out with remarkable efficiency. The galaxy M84 was the subject of an early STIS search for a black hole. Figure 4.19 shows a WF/PC-2 image of the central region of the galaxy (with the location of the STIS entrance slit over the site of the black hole marked) and the resulting STIS spectrum at each point along the slit. The rotational motion of stars and gas is determined from the Doppler shift of emission lines from the gas and is coded in the resulting spectra by color. If there were no rotation, there would be no Doppler shifts, and the spectrum would be a straight, vertical line of green and yellow picture elements. However, the speed of rotation rapidly increases as the central disk is approached with the upper part (blue) moving toward us and the lower part (red) moving away from us at 400 kilometers per second. From these speeds, the mass of the central black hole is determined to be about one and one-half billion Suns. Using this type of approach with STIS, the detection of massive central black holes in active galaxies could soon become more routine than it already is!

Additional research finds that nearly all galaxies, even ones that appear 'normal', i.e. those with none of the usual signs of activity such as jets or a central disk, may have a central black hole. One study, led by Doug Richstone of the University of Michigan at Ann Arbor, utilized HST and ground-based telescopes in Hawaii. From this work, some generalized deductions about supermassive black holes are starting to coalesce. First, supermassive black holes are very common – nearly every large galaxy has one. This means that formation of supermassive black holes is a common process in galaxy evolution. Secondly, the mass of the black hole is proportional to the mass of the host galaxy. Thus, a galaxy that is twice the mass of another would have a black hole twice as massive. This result suggests that production and growth of the black hole are integral parts of galaxy formation and evolution. Finally, the number and

| 40 000 light years | 2000 light years | 40 light years |

Figure 4.20. Multiple views of the Andromeda Galaxy, M31. The images at left and center are ground-based views of the galaxy and its core, respectively. The WF/PC-1 (in PC mode) image (at right) clearly shows a double nucleus. See the text for discussion. *(T.R. Lauer, National Optical Astronomy Observatories; NASA)*

masses of the black holes are consistent with the requirements to power the quasars, as we will explore later in this chapter. This poses an interesting question as we try to relate galaxies to the brightly burning quasars – one that we hope to answer in the near future: could all or most galaxies have burned as brightly as quasars in the past? The answer could be an exciting one, no matter what it is.

M31

One of the most interesting galaxies studied by HST to date is right in our neighborhood – the Andromeda Galaxy (M31). According to astronomer Sandra Faber, Andromeda is an intriguing place to study, and no one is quite sure what is happening there.

Faber and her colleagues found a double nucleus at the heart of Andromeda. Double nuclei may be important in the evolution of galaxies, and also may be important in understanding quasars. Figure 4.20 shows M31 with scales from 40 000 light years for the ground-based view of the entire galaxy to 40 light years for the HST view of the nuclear region.

At this relatively early stage, Faber and her team have at least two working explanations of the situation. One is that the brighter component, apparently composed of densely packed bright stars, is a remnant of a collision between M31 and another galaxy which it cannibalized. A black hole probably lies at the center of M31, and it would devour the other core in just a few hundred thousand years, according to Kitt Peak National Observatory astronomer Tod Lauer: 'This is very short in cosmic time. We would have to be looking at the galaxy at a very special time to see it now.'

The other explanation for the peculiar appearance of the M31 core is that there is a band of absorption by dust, which creates the illusion of two peaks of light. Whatever is really happening there, the central region of M31 is much more complex than previously thought. As Faber puts it, 'I love it. This is typical astronomy. You get these little clues about something and you try to put it all together. It's wonderful.'

Only 25 years ago, black holes were considered to be something so strange and outside the boundaries of space and time that no one took their existence seriously. Yet, they spark the imagination. The idea of something that is so powerful that it sucks light into a gravitational well has passed into the daily usage as a reference to anything that sucks time, money, and energy. It is exciting to see that an object that was once the realm of science fiction has been proven to exist and that it could provide an explanation for so many of the most unusual things we have observed in the universe. HST's role in the search for these objects will surely prove the existence of many others embedded in the hearts of distant galaxies.

Quasars

In Chapter 5, we describe quasars as immensely bright beacons useful for probing the universe. Here we take a look at the engine 'under the hood' of quasars. Quasars were originally detected as radio sources, and, because these objects looked starlike, they were first dubbed 'quasi-stellar radio sources' – quasars.

Observationally, quasars exhibit large brightness variations, and some show jets; see the image of 3C 273 shown in Figure 4.21. They also appear to be immensely luminous, but here a bit of deduction is required. With an observed quasar brightness, we can determine the luminosity only if we can fix the distance. For years this was a major problem. Astronomer Maarten Schmidt of the California Institute of Technology solved the problem by noting that the spectra of quasars made sense if certain emission features had very large redshifts. The only accepted way to produce large redshifts is as part of the Hubble expansion of the universe. If the distances are assigned in this way, quasars, being very distant, therefore must be immensely bright.

Figure 4.21. A 4-meter Mayall Telescope (Kitt Peak National Observatory) photograph of the quasar 3C 273. This galaxy is 100 times more luminous than the brightest normal galaxy, and the jet measures some 150 000 light years in length. *(National Optical Astronomy Observatories)*

The fact that a quasar's brightness or luminosity can fluctuate on a timescale of a day or less implies that quasars are actually quite small. Simply put, large objects have no way to vary synchronously (i.e. all at the same time), and the shortest time by which an object can vary is the time it takes light to travel across the entire object. The observations confine the sizes of quasars to something quite small – perhaps no more than a few light days across (not much larger than the Solar System) for the light-emitting region.

The requirement of extraordinarily large energy production in a small volume, together with the existence of jets, point to the conclusion that supermassive black holes exist as the central engines for quasars. A black hole with a mass of 10 billion Suns ($10^9 M_S$) would have an event horizon 200 astronomical units across. The luminosity could be produced by only 1 solar mass of material falling into the black hole per year.

Now that we *think* we have a power source for quasars – a central, massive black hole – we need to press on and determine exactly how quasars work. How does the energy of matter falling into the black hole convert into the distinctive quasar spectrum we see from x-rays, to ultraviolet light, to visible light, to infrared radiation, to radio waves? Can we show that we really have a massive black hole in the core of a quasar by determining its mass? Recall from our discussion of M87 that we can only expect to establish the existence of massive black holes by gravitational effects on their surroundings. Therefore, if we had independent estimates of the mass of the black hole and the accretion mass rate, we could check the luminosity against the observations.

A comparison of nearby quasars with those existing 10 billion years ago implies that a typical quasar was then 100 times more luminous than those observed currently. This result implies that quasars – like galaxies – evolve over time. If we hypothesize higher mass accretion rates to make higher luminosities, then the masses of the central black holes in quasars must increase with time.

A fundamental understanding of quasar spectra would be a part of the process of answering questions about quasar evolution. Scientists using combined Faint Object Spectograph (FOS) and ground-based quasar spectra have seen broad emission lines in their data. In one interpretation, these lines are produced by clouds of gas in orbit around the black hole. If this is correct, the speeds of the clouds could be used to determine the mass of the black hole. The FOS spectra have observed wavelengths as low as about 1500 angstroms and provide access to the rich spectrum of the near ultraviolet, which includes lines at lower wavelengths redshifted into this region. When combined with ground-based spectra, the result is a series of spectra covering thousands of angstroms.

Fig 4.22. A rogue's gallery of quasar host galaxies imaged by WF/PC-2. *(J. Bahcall, Institute for Advanced Study, Princeton; M. Disney, University of Wales-Cardiff; NASA)*

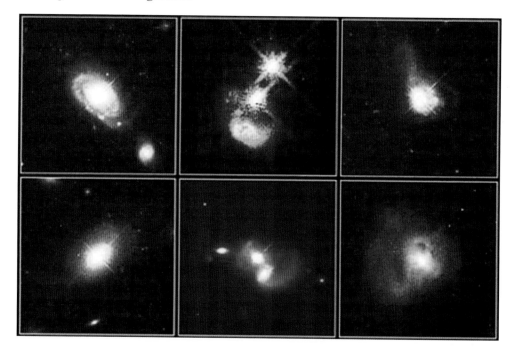

The imaging capabilities of WF/PC-2 have been used to probe the environment of quasars. Two teams, led by John Bahcall, at the Institute for Advanced Study in Princeton, and Mike Disney, at the University of Wales, have produced puzzling results, and a sample is shown in Figure 4.22. A brief summary illustrates the problems inherent in trying to determine the 'normal' home for a quasar.

In the top left panel, we see a quasar at the core of a normal spiral galaxy. The image in the top center panel shows a catastrophic collision between a quasar (central object) and another galaxy (below). A foreground star lies just above the quasar. At top right, a quasar has captured a tidal tail of dust and gas, presumably from a passing galaxy not in the image. In the bottom row, we see a quasar at the core of a normal elliptical galaxy at left, a quasar merging with a bright galaxy (just below the quasar) in the center panel; and finally, at bottom right, we see loops of glowing gas that appear to be the result of two galaxies merging. While these results certainly show that collisions between galaxies are important, many other quasar images show normal, undisturbed galaxies. In the survey by Bahcall's team, about half of the 20 quasars observed look undisturbed! This is a puzzling situation that does not lend itself to any sort of general conclusion about the home sites for quasars. Indeed, it prompted John Bahcall to remark: 'If we thought we had a complete theory of quasars before, now we know we don't. No coherent, single pattern of quasar behavior emerges. The basic assumption was that there was only one kind of host galaxy, or catastrophic event, which feeds a quasar. In reality we do not have a simple picture – we have a mess.'

The situation with quasars now is that, while astronomers have a great deal of really good information, a breakthrough on understanding the fundamental nature of quasars is still some way away. Quasars, it seems, are remarkably similar to cats. They behave in mysterious ways, do exactly as they please, and – so far – have frustrated our attempts to understand their behavior. Still, intensive work on theories to explain them continues, and as Robert Williams, past director of the Space Telescope Science Institute, has commented, 'Sometimes progress comes slow, sometimes serendipitous. Despite all the best efforts to really attack this in an analytical way, it may be some fluke that ultimately causes us to understand this.'

Collisions between galaxies

Arp 220

A major source of energy for quasars can be provided by galactic collisions, which can also cause enhanced star formation and produce spectacular images.

In our discussion of M31, we mentioned the possibility of galactic mergers as one way to explain some interesting structures seen in the central regions of galaxies. The 220th object in astronomer Halton Arp's *Atlas of Peculiar Galaxies* is an ultraluminous infrared galaxy located some 250 million light years away in the constellation Serpens, and the NICMOS image in Figure 4.23 shows wonderful structure forming as a result of a galactic collision.

The infrared image allows us to peer through most of the obscuration and see the collision of the two spiral galaxies at the heart of Arp 220. The crescent-shaped object (right) is the remnant of one of the colliding galaxies and contains about 1 billion stars. Part of this galaxy (below the crescent) is probably obscured by a very thick dust cloud. The core of the other galaxy is the bright, round object (to the left of the crescent-shaped object). The galactic cores are about 1200 light years apart and are believed to be orbiting around each other. The image was originally obtained in three infrared wavelength bands, but the colors in Figure 4.23 have been adjusted so that blue corresponds to the shorter wavelengths and red to the longer wavelengths – all in the infrared.

The collision of galaxies is now known to produce regions of dense dust and gas in which star formation – so-called 'starburst activity' – would proceed at a furious rate. The formation of very massive stars, such as the star Melnick 42 discussed in Chapter 3, would be expected in such regions.

Figure 4.23. This NICMOS image shows the central region of galaxy Arp 220. This is really two galaxies in collision, and it appears that the activity has produced a burst of star formation as a result. *(R. Thompson, University of Arizona, N. Scoville, California Institute of Technology; M. Rielke, G. Schneider, R.Thompson, University of Arizona; NASA)*

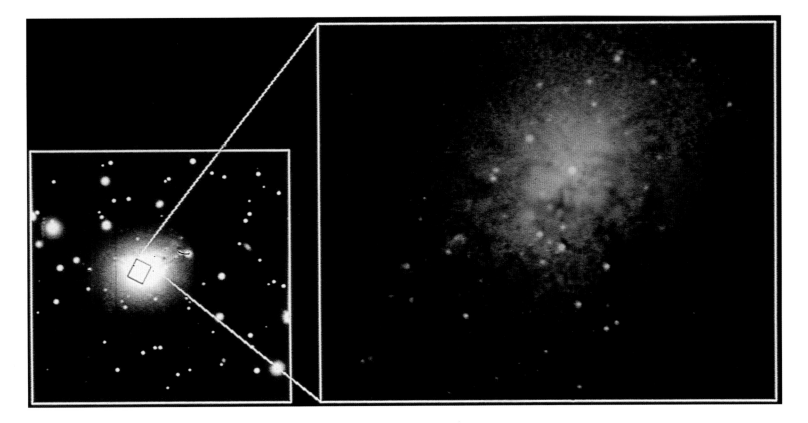

NGC 1275

At least one colliding galaxy is changing our viewpoint about galaxies and their associated globular clusters. Near the elliptical galaxy NGC 1275 – also called Perseus A – lies a collection of *young* globular clusters. Astronomer Sandra Faber and a team of scientists used WF/PC-1 to look at the core of the giant galaxy. 'It's the brightest galaxy within 200 million light years,' she said. 'It's at the center of a big cluster of galaxies, and in general this type of galaxy is thought to be product of a galactic merger. They're usually kind of bland-looking ellipticals and they look like they've been there for a long time. Perseus A is not like that. It's a big radio source; it's lumpy, full of dust and there's clearly a lot of gas around it. The spectrum is full of A-type starlight, which indicates there's been recent star formation.'

According to Faber, HST made two significant discoveries about this galaxy: 'The first was that we detected tidal arms on the image and that indicates that this galaxy did undergo a merger with something really massive, let's say within the last 300 million years,' she explained. 'So a lot of this activity could be traced to that. There's growing evidence that merging of gas-rich galaxies creates these intense starbursts, and the A-star light here. A-type stars live about 200 million years so it's perfectly consistent with a burst of star formation a few hundred million years ago triggered by a merger. That's point number one. Point number two is that we discovered a bunch of globular clusters. To be more precise, they are blue point-like sources very bright, a hundred times brighter than the brightest known globular cluster. We believe we've detected "proto-globular clusters"'.

The significance of this find is important to the understanding of the origin and evolution of globular clusters. Because these clusters usually contain very old stars, Faber and her colleagues expected to see reddish stars. Instead, the HST image of NGC 1275's central region shows individual star clusters that look distinctly blue, and therefore young.

The Cartwheel Galaxy

HST also captured the sight of a spectacular head-on collision between two galaxies located 500 million light years away in the southern hemisphere constellation Sculptor (Figure 4.25). The telescope resolved knots of star creation in the wake of the collision, and provided a new

Figure 4.24. Ground-based view (4-meter Mayall Telescope at Kitt Peak National Observatory) and WF/PC-1 image of the core of NGC 1275. The HST image reveals individual star clusters that appear as bright blue dots. These globular clusters contain young stars rather than old ones, which is the usual circumstance. *(J. Holtzman, University of California-Santa Cruz)*

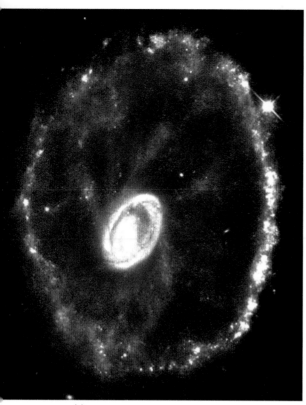

(a)

Figure 4.25. (a) (Above) A WF/PC-2 image showing detail in the Cartwheel Galaxy. The image shows the ring of newly formed stars that surround the central region. This star-formation activity is the result of a collision with another galaxy in the recent past. *(C. Struck and P. Appleton, Iowa State University; K. Borne, Hughes STX; R. Lucas, STScI; NASA).* **(b) (Right)** Although only an an artist's conception, this rendering of the comet-like clouds of gas in the Cartwheel Galaxy shows what might be happening there. The 'heads' are each a few hundred light years across, and the 'tails' are several thousand light years long. *(J. Gitlin, STScI; NASA)*

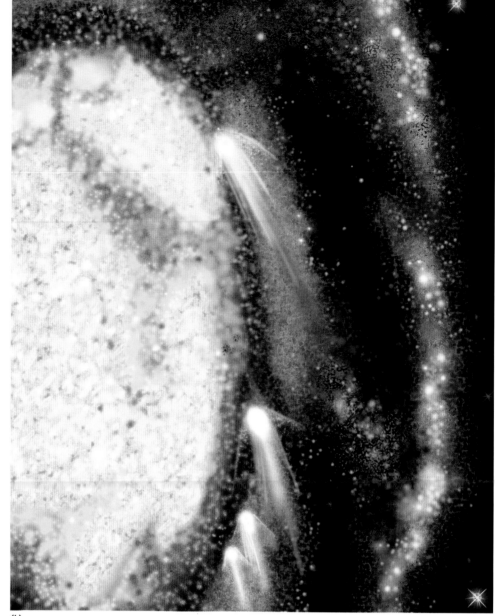

(b)

window on the formation of massive stars in large gas clouds. The large galaxy in the center of the ring was a normal spiral before its encounter with a smaller intruder galaxy. The ring is a direct result of another galaxy careering through the core of the host galaxy. The collision sent a ripple of energy out from the site, plowing gas and dust in front of it. A 'firestorm' of star creation resulted, and in other areas supernovae have been set off by the dynamics of the collision. The ring contains several billion new stars, evidence of starbirth on a grand scale and over an incredibly short period of time.

A close-up view of the central region of the Cartwheel reveals additional fascinating detail in the form of 'comet-like' structures that stretch across many light years. There is not yet a clear explanation for these objects, but these galactic 'comets' could be material splashed out in the collision and just now falling back into the galaxy.

Like archaeologists at a dig, astronomers must now figure out the sequence of events that occurred here 500 million years ago, and determine which galaxy plowed through the heart of the Cartwheel. Two nearby galaxies could be the interlopers, but some astronomers suggest that the galaxy involved in the collision has long since fled the scene of this 'cosmic traffic accident'.

Gamma-ray bursts

Whenever a concerted observational effort takes place in an unexplored region of the electromagnetic spectrum, new results are sure to follow. The situation for gamma rays is no exception. What are gamma rays? Recall the discussion in Chapter 1 on the electromagnetic spectrum. The light our eyes are sensitive to is the visible wavelength region. The photons of other spectral regions can have less energy or more energy. Infrared photons and radio wave photons have less energy, while ultraviolet and x-ray photons have more energy. We use the descriptive term 'gamma rays' for the most energetic photons known, and, essentially, we are talking about photons 100 000 times more energetic than visible light. Obviously, bursts involving photons with this high energy must come from a source that is indeed interesting. To detect these bursts, we use spacecraft above the Earth's atmosphere.

Cosmic gamma-ray bursts were discovered as part of the Vela satellite program monitoring the former Soviet Union for compliance with the nuclear test-ban treaty of 1963. Later, the results were declassified, and the published paper by Ian Strong and Ray Klebesadel, entitled 'Observations of Gamma-Ray Bursts of Cosmic Origin' appeared in the *Astrophysical Journal* for 1973. Since then, observations have been made by the spacecraft Ulysses, Compton Gamma Ray Observatory, X-Ray Timing Explorer, and Beppo-SAX. The accumulated observations have detected about three bursts per day. They come equally from all directions in the sky (an astronomer would say that they are isotropic). The nature of the bursts is highly variable. But, they rise to their maximum intensity in about one-tenth of a second. For an object to produce this radiation, it must be no farther across than the distance light travels in one-tenth of a second. This makes the source of a gamma-ray burster only a few tens of thousands of kilometers across! There has been controversy over the location of these bursters, but that may have been resolved with the discovery of a redshifted, $z \sim 1$ absorption line in the spectrum of a source that flared on May 8, 1997 (see below). These bursters are almost certainly at cosmological distances, some 3 to 10 billion light years away. At such distances, the luminosity of the source must be in the range 10^{51} to 10^{52} ergs. To give you an idea of the energy, consider that the Sun at present emits about 4×10^{33} ergs per second. If we choose a gamma-ray energy of 4×10^{51} ergs, the Sun would have to burn at its present luminosity for 10^{18} seconds, or 30 billion years, to equal the amount emitted by the gamma-ray bursters in a fraction of a second. This is a tremendous amount of energy!

The production of this huge amount of energy in a very small volume (cosmologically speaking) means that we are dealing with major fireballs popping off in the universe. The theoretical explanation for the bursts has proved to be a controversial one. The leading contender assigns the blame to a pair of neutron stars in a binary system. They spiral together, coalesce, form a black hole and release a titanic amount of energy to produce a cosmic fireball so bright that it can be seen anywhere in the universe. If that explanation seems a bit far-fetched, consider this alternative idea: binary systems composed of a stellar-mass black hole and a neutron star do the same thing as the neutron stars to achieve essentially the same result. Astronomers are not even close to an explanation for these objects, and it has been noted by at least one researcher that the number of papers on gamma-ray bursts exceeded the number of recorded bursts until 1996 (about 2500).

HST leaps into the fray by lending its potential to establish accurate positions and to follow the evolution of the sources, including the afterglow. This type of work is often a collaboration between gamma-ray, x-ray, and optical observations from spacecraft, and optical and radio observations from the ground. On February 28, 1997, a gamma-ray burst was detected by the Italian–Dutch satellite, Beppo-SAX. The burst lasted about 80 seconds, received the designation GRB 970228, and had its position determined to within a few arcminutes in the constellation Orion. Naturally, this detection set off a flurry of activity.

HST obtained images of the visible wavelength fireball on March 26 and April 7; the two images are combined in Figure 4.26. The arrow indicates the fireball. The feature to the lower right of the fireball (roughly E-shaped) is interpreted as being the host galaxy for the gamma-

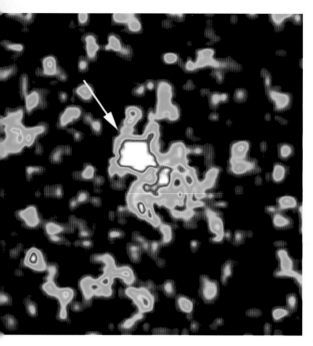

Figure 4.26. (Top left) This WF/PC-2 (in PC mode) image is a combination of two images taken on March 26 and April 7, 1997. The gamma-ray burst GRB970228 was first detected on February 28, 1997. *(K. Sahu, M. Livio, L. Petro, and D. Macchetto, STScI; NASA)*

Figure 4.27. (Top right) STIS image of the fireball from the May 8, 1997, gamma-ray burst. Note that no host galaxy is apparent here. *(E. Pian, ITESRC; A. Fruchter, STScI; NASA)*

ray burst. A comparison of the March 26 image with the April 7 image found that the fireball had faded, but that the neighboring source had not. This supports the idea of the other source in the image being the host galaxy.

The observation that the fireball is offset from the center of the galaxy rules out something associated with the galaxy's central region, such as a supermassive black hole, as the source. The offset is compatible with the idea of a collision of two neutron stars (or a neutron star and a stellar-mass black hole) in the disk of the galaxy as the source.

Another gamma-ray burst was detected by Beppo-SAX on May 8, 1997, and imaged by HST on June 2 (Figure 4.27). The fading fireball is seen in the center of the Space Telescope Imaging Spectograph image, but no host galaxy is visible. If there is one, it is much fainter than the Milky Way. If one of the faint galaxies shown a few arcseconds from the fireball is the host, then the gamma-ray burst originates in an extended halo, rather than from the galaxy's disk.

All in all, the HST observations of gamma-ray burst GRB 970508 produce quite the opposite conclusions as the observations of GRB 970228. For now, HST seems to be following in the gamma-ray burst tradition of great, new observations that produce much puzzlement but little new understanding of these energetic events.

The high resolution and light-gathering power of the HST have provided scientists with powerful ways to study galaxies – particularly in the ultraviolet. As HST continues in its mission, new cameras and spectrographs will expand its ability to capture unique characteristics of galaxies and will give astronomers a clearer view of these giant stellar cities. Studies of the long-scale evolution of galaxies have implications for the structure of the universe. Ultimately, we can use what we learn from the evolution of clusters of galaxies to understand the origin and evolution of the universe – and it is to this that we turn our attention next.

Cosmology

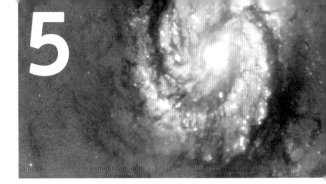

At last we come to cosmology – the sometimes mind-boggling, always captivating study of the entire universe – its origin, evolution, structure, and ultimate fate. There are some profound ideas in cosmology, notions of causality and eternity that inevitably drift across one's mind while viewing the stars on a dark, clear night. The scientific questions at the heart of cosmology seem to be grounded in musings of a more philosophical and sometimes even a religious nature. How far away is it all? Where exactly is the 'edge' of the universe? Can we see it? And when we do, what will we find? To answer these questions, we need to answer another one – how old is the universe? which begs yet another question – will the universe just keep getting bigger, making more stars and galaxies? Or does something else happen to it?

Historically, the human view of the universe has been one of ever-expanding horizons. For millennia people cast about for explanations of the origin and evolution of the Earth, and the genesis of anything beyond the Earth was simply a mystery. Until relatively recently, our view of the universe placed the Earth at the center, with everything else – the Sun, Moon, planets, and stars – revolving around it. Today our 'world-view' of the universe is changing again. It is as if we are standing at the foot of a cosmic ladder anchored on the Earth, looking up to a very distant top rung. We have no way of knowing exactly how far away it is yet, but we do have the means to climb the ladder. We know that our place in the universe is not central, but we are still trying to figure out just where we are in the grand scheme of things, and how far away we are from everything else. Some of this 'desire to know' is driven by simple curiosity: if we know the physical limits of our universe, we can approach an understanding of its origin and evolution.

Telescopes, cameras, spectrographs, photometers – all of our astronomy hardware gives us a leg up on the distance ladder by making it possible to see ever more distant galaxies and quasars. Orbiting sensors such as the Cosmic Background Explorer magnified for us the last whispers of the Big Bang, echoing across the cosmos. The International Ultraviolet Explorer opened a window to a universe of frenetic activity. We have taken the step up to the first rung of the ladder, but there is still a long way to go.

Why this interest in cosmic distances? Simply put, if astronomers can determine accurate distances to faraway objects in the universe, they will have a good start toward determining its age, origin, and evolution. Using the Hubble Space Telescope, astronomers have found some of the most distant galaxies in the universe, which are some of the youngest structures in the cosmos. Now the fun begins: determining *exactly* how far away these objects lie from us.

The Hubble Space Telescope's contribution to determining cosmic distances is best described as a work in progress. During its first seven years, Hubble's work was a veritable laundry list of cosmologically interesting observations:

- an accurate determination of the deuterium-to-hydrogen ratio in the interstellar medium;
- the detection of more than a dozen intergalactic hydrogen clouds within a million light years of the Milky Way galaxy;
- the detection of Cepheid variables in other galaxies;
- converging on a new value for the Hubble Constant, from highly accurate measurements of the distance to galaxies in the Virgo and Fornax clusters, using Cepheid variable stars.

The most beautiful thing we can experience is the mysterious. It is the source of all true art and science.

Albert Einstein

We had the sky up there, all speckled with stars, and we used to lay on our backs and look up at them, and discuss about whether they were made, or only just happened.

Mark Twain

Why are these topics so interesting to cosmologists? One way to answer this is to examine some of the basic concepts and assumptions of cosmology. The two main concerns are distance and age.

First, much of Hubble's work is involved with determining accurate distances to faraway objects that are difficult or impossible to see with ground-based instruments. As we have discussed elsewhere, its position above the atmosphere is ideal for ultraviolet spectroscopy and high-resolution infrared studies, which are important tools in cosmological research. However, the work is not as straightforward as targeting dim objects and making long exposures, or taking a spectrum of a distant object, measuring its redshift, and then declaring that it lies billions of light years away. Many discussions of cosmology do include a statement that an object is probably at a certain distance. While the statement may be true, it hides the struggle involved in pinning down distances to anything but the nearest objects.

The other side of the cosmological coin involves determining ages of distant objects, in the hopes of discovering just how old the universe might be. Certainly distances are one way of zeroing in on an age for the cosmos, but we can make other types of observations that bear directly on the process of dating the universe. For example, using the technique of radioactive dating, geologists have determined that the Earth, the Moon, and the meteorites are around 4.5 billion years old. The Sun's age is taken to be approximately 5 billion years. Using techniques discussed in Chapter 1, we can then turn out attention to studying stars outside our own system.

Analysis of starlight and our knowledge of stellar evolution gives us the age of other stars. Once we characterize them according to their size and luminosity, we find that their ages range from a few thousand years – for the very youngest newborns – to at least 10 billion years for the oldest red giants. Some very old stars lie in globular clusters near the core of the Milky Way, and *appear* to be 13–15 billion years old. Since the globulars are thought to have formed at approximately the same time as the galaxy, we know that the Milky Way Galaxy is at least the same age. The Milky Way is about the same age as all the other galaxies. Paradoxically, other galaxies *appear* to be much younger than the Milky Way because they are so distant. Some of the distant galaxies we have seen emitted their light when the universe was only half the age it is now. Farther away in space and further back in time lie other galaxies that must have formed when the universe was only 20 per cent of its present age. Theoretically, the further back we look, the younger are the galaxies we see. Because the universe is expanding, finding the limit of the observable universe may well prove to be impossible. But, we can continue to look across the cosmos, finding ever younger, ever more distant structures.

Cosmological distances

The astronomical distance scale starts right here on Earth. We look out and see the Sun, Moon, and planets, and wonder how far away they are. Celestial mechanics and simple observations determine the relative spacing of the planets, and radar observations determine the absolute scale very accurately. So, for example, the distance from the Earth to the Sun is 149 million kilometers. Light takes eight minutes to traverse that distance, making the Sun 8 light minutes away. The moon is 380 000 kilometers from Earth, and light takes 1.27 seconds to make the round trip. So you could say that the moon is 1.27 light seconds away. Mars is around 228 million kilometers from the Sun, or 13 light minutes away. The distant outer planet Pluto is 5900 million kilometers from the Sun, or 5.5 light hours away.

These distances are relatively easy to manage, but when we start to look at nearby stars and galaxies, such as Proxima Centauri, which lies 4.1×10^{13} kilometers away, or the Large Magellanic Cloud, which lies 1.61×10^{18} kilometers away, you can see that the distance numbers start getting out of hand. Astronomers have adopted several shorthand 'standard' references for large distances. One method is to use the astronomical unit (or AU) – the average distance between the Earth and the Sun, which is 1.5×10^8 kilometers. Thus, Mars lies at 1.5 astronomical units and Pluto is about 40 astronomical units away. However, that system also breaks down when we move to the stars and galaxies. The closest star to our Sun, Proxima Centauri, is more than 273 000 astronomical units and the Magellanic Clouds are 10^{10} astronomical units away. Obviously, we need to use larger units for more distant objects. It is often easier to use a unit that gives us another way of understanding distance – the distance light travels in a year (9.5×10^{12} kilometers), more commonly known as a light year. Thus, Proxima Centauri is 4.3 light years away, and the Large Magellanic Cloud is just over 170 000 light years distant.

We can also use the parsec, a unit based on angular measure which is 206 265 astronomical units, or 3.26 light years. Thus, the distance to Proxima Centauri becomes 1.3 parsecs. The Large Magellanic Cloud is 52 000 parsecs away or 52 kiloparsecs. For even larger distances, the system once again gets unwieldy, so we speak in terms of millions of parsecs, or megaparsecs. These are huge numbers describing incredible distances.

Table 5.1 *A comparison of distances*

Distance unit	Kilometers	Light-travel time
light second	3×10^5	1 second
light minute	1.8×10^7	1 minute
astronomical unit	1.5×10^8	8 minutes
light year	9.5×10^{12}	1 year
parsec (pc)	3.1×10^{13}	3.3 years
kiloparsec (kpc)	3.1×10^{16}	3.3 thousand years
megaparsec (Mpc)	3.1×10^{19}	3.3 million years

People often wonder just how astronomers *know* these distances. The answer is that they use a variety of methods, both direct and indirect, to measure distances to objects in the universe.

For stars relatively close to us, one technique of distance determination involves using a combination of parallaxes (the apparent shift in the position of the star when viewed from Earth over the course of a year) and a more esoteric technique (measurements of moving clusters) to measure distances to clusters of stars. If stars in other galaxies and globular clusters can be classified according to the Hertzsprung–Russell diagram, the main sequence itself can become a tool for determining distances.

In 1924, Edwin Hubble determined the distance to a nearby, naked-eye object he knew as the Andromeda 'Nebula'. His tool was a type of star called a Cepheid variable – named after the fourth-brightest star in the constellation Cepheus, itself a variable. Cepheids have a very interesting and useful characteristic: their intrinsic brightness changes over periods of time ranging from 1 to 50 days. These brightness changes obey a period–luminosity law developed by astronomer Henrietta Leavitt in the early part of the 20th century: brighter stars vary over shorter periods of time. Thus, if we measure the period of a Cepheid variable (that is, the length of time it takes to go from maximum brightness to minimum brightness, and then back again), we can determine its intrinsic brightness. Because the intensity of a star's light fades as a function of its distance from us, comparing the intrinsic brightness of the Cepheid with its apparent brightness gives us a distance. Edwin Hubble applied this method to Cepheids in Andromeda and thus estimated the distance to that galaxy.

Cepheids are *standard candles* – objects in the universe that astronomers use to determine distances. Think of standard candles in the following way: imagine that you are standing at one end of a large room looking at a few light sources against the wall at the opposite end of the room. If all of the sources are 100-watt light bulbs, and they are all at the same distance, they should all look equally bright. If some bulbs are closer than others, they will look brighter, but they will not actually *be* intrinsically brighter than the more distant ones. They will always shine as 100-watt light bulbs no matter how far away they are. What astronomers look for are objects that have the same *intrinsic brightness* – in essense, cosmic 100-watt light bulbs. Some might be a few light years away and look bright, and others might be a few megaparsecs away and look dim. Because they have the same intrinsic brightness, however, their function as distance indicators is important. Cepheid variables are one example of this type of source. And, if we can find Cepheids in galaxies, or other candles such as the variable type RR Lyrae stars in globular clusters, then we can determine accurate distances to them.

Armed with these observations of standard candles, astronomers can accurately determine the distances to nearby galaxies such as the Large and Small Magellanic Clouds, the Andromeda Galaxy (M31), and its companion, M32. Extension of the distance scale to a nearby cluster of galaxies can also be done by measuring the magnitude of the brightest stars. Actually, an entire galaxy can be a standard candle, as can the brightest galaxy in a cluster, or even a specific type of supernova, which itself can rival an entire galaxy in brightness.

Of course, all these methods may seem deceptively easy, but there is a catch. Stars and galaxies are not hanging perfectly still in space, waiting for us to measure them. Everything in the universe is in motion. This motion affects our measurements, and so when determining distances, we must factor in another number. That number is called the Hubble Constant – or, as cosmologists like to refer to it – H_0 or 'H-naught'. H_0 indicates the rate at which galaxies recede from each other as a function of the separation between them. It is tied to the spectral

Figure 5.1. The schematic illustration of the steps necessary to determine a reliable value of the Hubble Constant, H_0, using the Eiffel Tower as a framework. This diagram presumes that the Cepheid variables are already calibrated as distance indicators. Then, Cepheids are used to determine the distances to relatively nearby clusters of galaxies: Virgo, Coma, Fornax, and Leo I. The distances can also be calculated using the so-called 'secondary indicators' that form the feet of the Tower. The four secondary indicators marked are representatives for all such indicators: TF is the Tully–Fisher relationship which gives the galaxy's luminosity from the width of the hydrogen radio-emission line at 21 cm; PNLF, the Planetary Nebula Luminosity Function, gives distances if the brightest planetary nebulae all have the same brightness; SBF, the Surface Brightness Fluctuation method, derives the luminosity from fluctuations in brightness across elliptical galaxies; TRGB, the Tip of the Red Giant Branch in the H-R diagram (see Chapter 3), gives distances if the position is universal. Iteration between distances determined from Cepheids and secondary indicators calibrates the latter. These indicators, including supernovae (which are very bright, but not observable in all galaxies), extend knowledge of distances farther than Cepheids can be used. The goal is to get beyond the nearby clusters (where gravitational attractions can distort the speeds) to the undisturbed 'far-field flow'. Then, we will have an accurate, solid value for H_0. *(Adapted with permission by G. Dinderman, Sky Publishing Corporation, from a diagram by Wendy Freedman and Barry Madore, IPAC and Caltech)*

redshift of these galaxies (which we will discuss shortly). From the cosmological perspective, accurate distances are the key to the Hubble Constant.

The elegant complexity of determining H_0 has been beautifully illustrated by astronomer Wendy Freedman. To cap off a discussion of her research in determining H_0, she presented her vision of the Hubble Constant distance ladder to a group of scientists at the Second Hubble Space Telescope Science conference, held in 1995 in Paris, France. Her Eiffel Tower diagram (Figure 5.1) clearly shows the interdependent steps scientists need to take to construct a reasonable value for H_0.

The Hubble Constant, along with a number of other factors that need to be taken into account when determining the age of the universe, plays an important role in creating a theoretical construct of the evolution of the universe called the 'Standard Model'. Think of this as the cosmological equivalent of a blueprint for the universe. As such, the Standard Model has inspired much of HST's cosmology work.

The expanding universe and the Standard Model

In the early part of the 20th century, as techniques to estimate distances to faraway stars and outlying galaxies were developed, Edwin Hubble built on another fundamental discovery. The spectra of galaxies generally showed redshifts; i.e. the spectral lines appeared to be shifted to the red or longer wavelengths. He reasoned that, if this is interpreted as a Doppler effect, it means that the galaxies are all moving away from us. Moreover, a comparison of distances and redshifts showed that the farthest galaxies appear to be moving away from us the fastest. In other words, the farther away the galaxy, the larger the redshift and, thus, the higher its speed of recession.

From this observation, the idea of the 'expanding universe' was established. However, it is important to remember that, even though all the galaxies are receding from us, that does not mean that the Earth, or the Sun, is at the center of the universe. To understand this apparent paradox of thought, consider the most popular depiction of an expanding cosmos: a balloon on which we have painted dots to represent galaxies. If we blow up the balloon, its surface expands and *all* the dots recede from each other. Thus, there is no central dot. In addition, each dot 'sees' every other dot receding, and those dots farthest away recede fastest. The correct view is that there is no center or preferred location of the universe. Specifically, the location of the Earth or the Sun, or anything else, is not unique in any way.

With the idea of an expanding universe and some reliable ways to measure distances, astronomers finally come face to face with the big question in cosmology: *how old is the universe?* This is a great question to debate at cocktail parties or on the Internet newsgroups. Of course, cosmologists (the astronomers who study the origin and evolution of the universe) do not really ask the question in that way. They are more likely to frame their questions in terms of things they can measure, such as the density of the universe, which influences the expansion rate of the universe. So they ask their own types of cosmological questions: How long ago did the universe begin? Is the universe expanding? Is the expansion rate the same everywhere in the universe? How does the density of the universe influence its expansion rate? What is the critical density of the universe? How dense does it have to be if it is to expand forever? What density do we need if we want the universe to stop expanding forever? And so on.

As you might expect, cosmologists, being scientists, assign numbers and letters to the key ideas in these questions. If we know the distances between galaxies and the speed at which they are moving apart, then theoretically we can calculate the time at which all the galaxies were together. This brings us back to the crucial number we need to do this calculation: H_0, which gives the current expansion rate of the universe in units of kilometers per second per megaparsec.

Next, we need to estimate the age of the universe – a variable we can call 'Time Zero' or label as T_0. You could also think of this as 'when time began', sort of like starting a cosmic stopwatch. This number is directly related to the Hubble Constant for reasons that we will see shortly. For now, consider some possible numbers to 'plug into' T_0. For the sake of argument, if we use numbers somewhere in the range of 10 to 20 billion years, we would get Hubble Constants of 100 kilometers per second per megaparsec and 50 kilometers per second per megaparsec, respectively.

The higher the value of H_0, the lower the age of the universe; i.e. the faster the universe expands, the less time it will have taken to reach its current size. We should realize that estimating the universe's age in this way leads to an upper limit (or overestimate) because the effects of gravity could have slowed the rate of expansion. The uncertainty in T_0, affected principally by uncertainties in distances and density estimates, illustrates that there is still a lot of work that needs to be done when it comes to determining the true age of the universe. If the density is not accurately determined, then we will have no way of fixing T_0 correctly.

There are other properties of the expanding universe: Q_0, which is a deceleration parameter (the rate at which the expansion of the universe slows down). This can be used to test the possibility that the exact Hubble relationship may differ for galaxies at different distances. There is also something called ρ_0 (pronounced 'rho naught') – the critical density of the universe, which would be exactly the average density of matter needed to stop the expansion of the universe. The density is usually discussed in terms of the current density, which we divide by ρ_0. This gives us a number called Ω_0 (omega naught). If Ω_0 is greater than 1, the universe will stop

Figure 5.2. What are 'open' and 'closed' universes? An open universe ($\Omega_0 < 1$; left) expands forever because it does not contain enough mass to slow the expansion of the universe. A closed universe ($\Omega_0 > 1$; right) has enough mass to stop the expansion, and, ultimately, the universe collapses (bottom right). A universe with exactly the right amount of mass ($\Omega_0 = 1$) is balanced between these two alternatives and expands forever, but at an ever-slowing rate. *(STScI)*

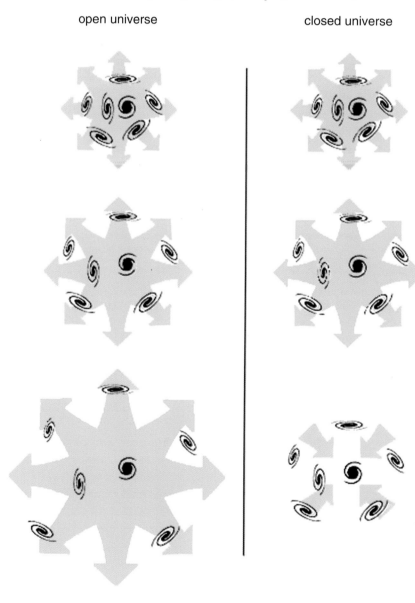

open universe   closed universe

expanding and will collapse back on itself in something called the 'Big Crunch'. This state of affairs would give us a 'closed' universe. If Ω_0 is less than 1, the universe will continue to expand, giving us an 'open' universe.

In addition, it is clear that if gravity has slowed things down to the present rate of expansion, the past rate of expansion had to be faster. Thus, the question of the universe's ultimate fate – indefinite expansion versus collapse and possible renewal – is one of the most fiercely debated in all of cosmology. Progress in this area of cosmology has refined the Standard Model of the beginning of the universe – a model which says, simply: in the beginning, things happened very quickly.

The birth of our universe is often called the Big Bang, which was followed immediately by an extremely dense, hot universe called the *Primordial Fireball*. Theoretical studies by astrophysicists such as George Gamow and his collaborators in 1948 led to the conclusion that a remnant of the Primordial Fireball existed, probably in the form of a microwave background radiation permeating all of the universe. This radiation was discovered in 1965 by Arno A. Penzias and Robert W. Wilson.

The universe as a Primordial Fireball initially (about 10^{-43} seconds after the Big Bang) consisted of energetic, elementary particles and energetic photons. This is the domain where cosmology meets elementary particle physics. At about 10^{-34} to 10^{-30} seconds, the universe is thought to have passed through a stage of very rapid expansion known as 'inflation'. Around 10^{-6} seconds, the elementary particles (called quarks) combined to form neutrons and protons. By about 1000 seconds after the Big Bang, the universe had cooled, and the formation of major elements in the Primordial Fireball was complete. These elements were hydrogen, deuterium, helium, and lithium, and they formed in proportions determined by the value of Ω_0.

At some point, the matter and the photons (light) no longer interacted, and they evolved separately. The matter simply cooled as the universe expanded, as did the electromagnetic radiation, from 10^4 kelvin to a current, almost-uniform background radiation at 3 kelvin. This has been observed in detail by the Cosmic Background Explorer (COBE) satellite, and fine structure has been detected. A summary of the Big Bang picture is shown in Figure 5.3.

A skeptical reader may wonder how *anyone* can talk with any certainty about events that occurred over 10 billion years ago, 10^{-43} seconds after the Big Bang. The answer is that events of long, long ago leave unmistakable traces even today. The 3 kelvin echo of the Big Bang is one such trace, and simple observations of the abundance of helium relative to hydrogen in the interstellar medium, and on the surfaces of many stars, provide another indirect, but valuable, way to check on conditions in the early universe. This, in turn, allows us to start fixing some upper and lower limits on the amount of matter in the universe. When those limits are known, we can then put out some useful values for Ω_0.

Knowledge of fine structure is particularly important in understanding the evolution of matter itself. Simply put, fine structure implies the existence of denser regions in the universe, which makes it easier to form matter into galaxies.

The organization of galaxies into larger units occurs in two stages: clusters and superclusters. Examples are the Virgo Cluster and the Hercules Cluster (Figure 5.4). Galaxy clusters are important both in cosmology and in the study of the evolution of galaxies (see Chapter 4). The biggest known structures are immense filaments and sheets of galactic clusters and superclusters. Some have picturesque names, such as the 'Great Wall'. Undoubtedly, these large-scale structures are the remnants of the fine structure seen in the COBE results, which mirror fluctuations in the very early universe that were preserved by its inflationary expansion. Once protogalaxies had formed in these structures, the stage was set for the condensation of gas into stars. This was followed by stellar evolution, with its implications for the interstellar medium, the formation of planets, and the development of life.

Not all astronomers agree with the Standard Model of the universe we have just discussed. Some scientists offer competing theories for the origin and evolution of the universe: for example, the idea of the steady-state universe proposed by Herman Bondi, Thomas Gold, and Fred Hoyle.

Figure 5.3. A summary diagram of the origin and evolution of the universe consistent with the COBE observations. Just after the Big Bang, small fluctuations existed in the universe, which, after rapid expansion, became the fluctuations seen by COBE. These were indicative of favorable conditions for galaxy formation. Finally, we have the current universe some 15 billion years after the Big Bang. *(NASA-Goddard Space Flight Center)*

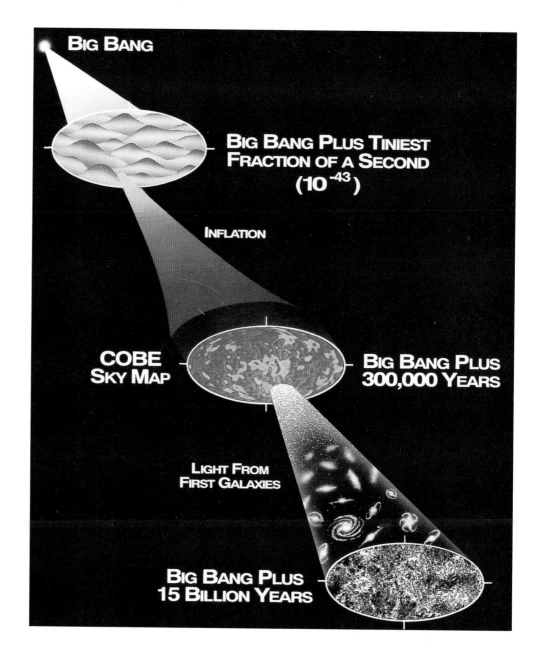

The steady-state situation is thought to be maintained by continually producing new matter (at a very low rate) to form galaxies that fill in the gaps produced by the expansion of the universe. This theory has produced an interesting/humorous side issue. The phrase 'Big Bang', now so inherent in our vocabulary, was actually coined by Sir Fred Hoyle to describe the 'Standard Model', possibly in an unflattering way. Hoyle still does not agree with the Big Bang picture, and an unsuccessful attempt to rename the Big Bang only served to illustrate that some questions are profound and require a long time to answer.

Dark matter

Regardless of the state of debate over which theory best describes the origin and evolution of the universe, there are still cosmic scores to settle. Basically, all the parameters mentioned above, H_0, T_0, Q_0, and Ω_0, need to be determined accurately. Astronomers are making good progress with H_0, but the density parameter may prove a tougher nut to crack. This is because there are many unknowns about how much matter exists in the universe. Specifically, the question of unseen 'dark matter' casts a shadow over any useful estimate for the density of the universe.

For decades, astronomers have known that much of the material in the universe is in the form of 'dark matter'. However, what is not known is how much there is. The existence of dark matter is deduced as follows: gravitational effects (on the motion of stars or galaxies, for example) imply a certain mass of material out there. However, searches for light-emitting sources fail to find enough to account for the estimated density of the universe. We know there is more matter out there, but we cannot see it, so we blame it on 'dark matter.'

Since we cannot directly observe dark matter, we have come up with many possibilities for what it might be, and some of the names are picturesque, for example MACHOs, or massive compact halo objects – very faint stars or Jupiter-like objects – which could be arranged in an essentially spherical cloud around galaxies. It is also possible that black holes are much more widespread than we suspect. Exotic particles could make up dark matter, and neutrinos could fill the bill if they have a small mass. Some evidence for this possibility exists, but, of course, this remains to be confirmed. WIMPs, or weakly interacting massive particles, might be good candidates for dark matter, and are predicted by theory. Theoretically predicted particles called axions are another possibility. These particles, if they exist, are remnants of the very dense, hot early universe that existed shortly after the Big Bang. The searches for these and other particles are underway, creating the relatively new field of particle astrophysics.

Figure 5.4. The Hercules Cluster of galaxies as photographed by the 4-meter Mayall Telescope of the Kitt Peak National Observatory. The core of the cluster contains less than 100 galaxies. *(National Optical Astronomy Observatories)*

HST and cosmology

HST's contribution to cosmology takes several approaches. First, there are the efforts to determine the density and makeup of the universe. These observations use quasars as candles to illuminate the intergalactic medium. In the same way, observations of relatively bright nearby stars have been used to determine the abundances of chemical elements in the interstellar medium. Studies of the amounts of deuterium and hydrogen relative to each other in the interstellar medium are critical to the process of determining the density of the universe. Stars near the Sun come in for their share of HST's cosmological studies because their distances can be measured astrometrically. Cepheid variable stars in the galaxy M100 have been studied to determine a preliminary value for the Hubble Constant using HST.

Secondly, turning its gaze toward the most distant reaches of the universe, HST is also peering farther into space, further back in time, to some of the most distant galaxies known. Finally, we will round out our discussion of HST's contribution to cosmology by looking at the work HST observers are doing using the phenomenon of gravitational lensing to test Einstein's Theory of General Relativity.

Lyman-alpha forest in nearby quasars

Quasars, as we discussed in Chapter 4, are tantalizing astronomical 'animals' in the distant reaches of the universe. Aside from their own intrinsically interesting characteristics, it turns out that they make excellent 'candles' to illuminate the darkness that lies between them and us. Because quasars are the brightest objects known, we can see them from across a significant fraction of the universe. The spectra of quasars show numerous absorption lines (Figure 5.5) that are now known as the 'Lyman-alpha forest'. What is happening is that a small, but important, portion of the ultraviolet light from quasars is absorbed by something lying in between us and them in the intergalactic medium. That 'something' turns out to be clouds of predominantly hydrogen gas. These intervening clouds produce a forest of absorption lines.

The absorption lines are redshifted due to the expansion of the universe, and the amount of shift depends on the distance between us and the object we are studying. To understand this concept, imagine listening to a group of train whistles here on Earth. Close-by, slowly moving whistles will sound different than faraway, rapidly moving whistles, and this gives you an idea of how fast the trains are traveling and how far away they are.

The line producing the absorption from the hydrogen atom is called the Lyman-alpha line. Its wavelength in the laboratory is 1216 angstroms, a value about one-quarter of the wavelength of visible light. As we discussed briefly in chapter 4, when light is redshifted, astronomers use a quantity z to describe the amount of redshift. The value of z gives the shift in units of the original wavelength. For example, $z = 3$ means a shift of three times the original value or a new, total wavelength of 4864 angstroms; $z = 2$ produces a new wavelength of 3658 angstroms; $z = 1$ corresponds to a new wavelength of 2432 angstroms; and finally $z = 0$ means no shift at all or a wavelength of 1216 angstroms. Thus z becomes a shorthand way of thinking of the distance of an object by referring to its redshift.

Because wavelengths less than about 3200 angstroms are absorbed by the Earth's atmosphere, earlier ground-based studies could only deal with the larger values of z, essentially $z = 2$ and larger. For these values, the redshift moved the wavelength into the visible part of the spectrum, and thus it could be seen from the ground. To study the low values of z which are visible in ultraviolet spectra, astronomers needed an observatory above the Earth's atmosphere. Until the launch of HST, essentially nothing was known about the nearby or low-z Lyman-alpha clouds. Because there are fewer clouds in the line of sight as you go from $z = 3$ to $z = 2$, the expectation was that there would be very few lines from clouds for z close to zero.

A group of researchers, led by Ray Weymann of the Observatories of the Carnegie Institution of Washington, used the quasar 3C 273 to study the line of sight between us and the quasar. 'For several years now, people have known that there were these clouds of

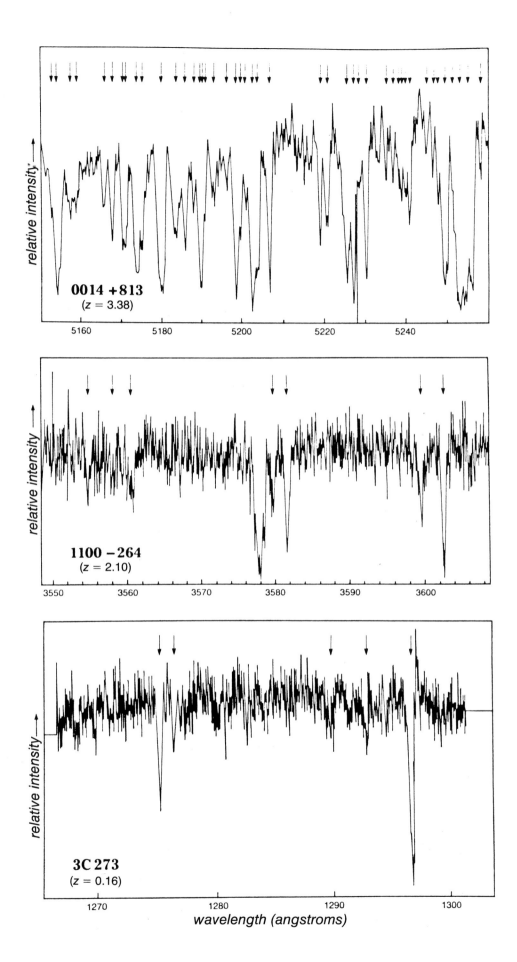

Figure 5.5. The Lyman-alpha forest for galaxies at $z = 3.38$, 2.10, and 0.16. The latter is the GHRS observation of 3C 273. Individual Lyman-alpha lines are depicted by the arrows. The large decline in numbers between $z = 3.38$ and $z = 2.10$ led to the expectation of very few lines at $z = 0.16$. This was not the case; see the text for discussion. *(Simon L. Morris, Observatories of the Carnegie Institution of Washington; Sky Publishing, Inc.)*

hydrogen between us and distant quasars,' he said. The result of that work is that there turned out to be more clouds between us and this nearby quasar than anyone had anticipated.

Weymann used the Goddard High Resolution Spectograph, GHRS, to observe this nearby quasar, which at $z = 0.16$ lies 430 Mpc away. The results are shown in Figure 5.5 in which the Lyman-alpha forest lines are marked; the other absorption lines are produced by the interstellar gas in our own galaxy. The result is that there is no significant difference in the number of lines from $z = 2.1$ to $z = 0.16$. Thus, the population of Lyman-alpha absorbing clouds is apparently unchanged over this interval.

What this means is that the clouds simply exist – and there may be neither creation nor destruction of clouds. Or, if some of these processes are active, the net effect is still no change. This is a great example of the importance of making observations, even when we think we know what we are looking for – in this case, what was expected was no Lyman-alpha forest at low z, and few interstellar clouds. These results have also been confirmed by FOS spectra of 3C 273 and other nearby quasars.

The origin of these clouds is interesting to contemplate. Are they clouds associated with galaxies, perhaps a residue of the process of galaxy formation? Or are they unrelated to galaxies, being held together by a very hot intergalactic gas? The preliminary feeling is that the clouds could well be associated with galaxies, and that more study is needed to explore this idea. As Weymann wrote in a 1994 paper describing his work:

> While the luxuriant growth of the very high redshift Lyman-alpha forest has given way to a rather arid savannah at low redshifts, their continued study is proving of importance for addressing the fundamental question of the origin and evolution of galaxies and cluster of galaxies.

This is not the last we will see of these hydrogen clouds. University of California at San Diego astronomer Margaret Burbidge, who has been trying to understand the physics of quasars and their environments for years, explained their importance. 'There's been a lot of interest in trying to see if there are other chemical elements in these hydrogen clouds that are producing absorptions. Are they primordial clouds of gas that form galaxies? Are they forming galaxies? Are they something blown off other objects? There is a lot of fascination out there, and we do not understand them very well. Apart from the interest in what we get from the quasar itself, we found that there were some absorption features in Lyman-alpha at different redshifts, lower redshifts than the quasar, and this has been one of the major discoveries of the Space Telescope.'

The deuterium-to-hydrogen ratio in the local interstellar medium

The abundances of elements has been a common thread in the studies carried out by HST. As we mentioned earlier, the lighter elements hydrogen, deuterium, helium, and lithium were created in the Primordial Fireball, with their proportions depending on the density of the fireball. That density also determined the value Ω_0. The amount of deuterium created (measured by the ratio of deuterium to hydrogen, D/H) is small and quite sensitive to this density, but the amount of the other elements is relatively insensitive.

The D/H value can be determined from observations of stars near the Sun because their light can serve as a probe of the local interstellar medium. However, deuterium has its fundamental absorption line (Lyman-alpha again) displaced some 0.3 angstrom from the hydrogen line. Because there is so much more hydrogen than deuterium in the universe, the hydrogen Lyman-alpha line would completely swamp the deuterium Lyman-alpha line unless a nearby star was used. So, the GHRS was used to measure light from the star Capella, which lies some 12.5 parsecs away from the Earth.

The spectrum for these measurements is shown in Figure 5.6, in which the deuterium Lyman-alpha absorption line is clearly visible. A detailed analysis has been carried out by a team of scientists led by Jeffrey Linsky of the University of Colorado and National Institute of Standards and Technology's Joint Institute for Laboratory Astrophysics.

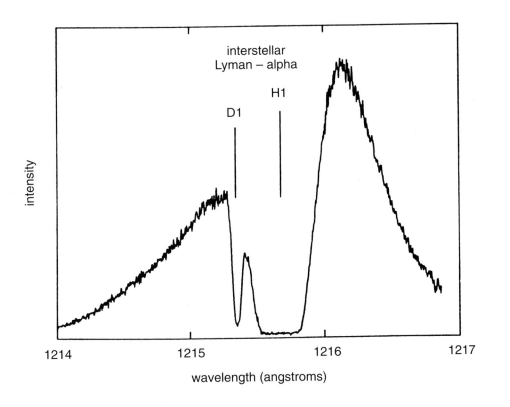

Figure 5.6. Spectrum of Capella taken by the GHRS in the highest resolution mode. The features of interest are the absorption lines caused by hydrogen atoms (marked H) and deuterium atoms (marked D) in the interstellar medium between Earth and Capella. Analysis of the spectrum determines the D to H ratio. *(J. Linsky, Joint Institute for Laboratory Astrophysics and University of Colorado)*

Linsky found the value for the abundance ratio as $D/H = 1.65 \, (+0.07, -0.18) \times 10^{-5}$. This is an extremely small number – an amount Linsky likes to call 'the Amazing Trace'. There had been attempts to measure this ratio previously, and this new measurement is close to the average of the previous values. However, the uncertainty in the GHRS value is much less, and it helps place a tighter constraint on the density of the universe.

Of course, this approach can be extended to other stars in the solar neighborhood, including some white dwarfs. This, in fact, has been done by several observers, and it turns out that new measurements made in the local interstellar medium are consistent with that of Capella. However, we need to bear in mind that we have sampled only the region near the Sun, and that these values need to be reconciled with other D/H measurements, particularly in quasars. Then we could understand the evolution of the D/H value since the Big Bang.

To the best of our knowledge, all the deuterium in the universe was created in the Primordial Fireball between 100 and 1000 seconds after the Big Bang. There are no other processes that create deuterium in the universe, but there are certainly ways to destroy it. The nuclear reactions in stars eat up deuterium, and much of the interstellar gas in our galaxy has been through several stars (see the discussion on stellar evolution in Chapter 3). Thus, the measured value needs to be increased to account for the deuterium destroyed in stars. This gives us an estimate for the primordial value of 1.5 to 3.0 times higher than today's value. What this means is that a higher value of D/H corresponds to a lower density of the universe.

The information in Figure 5.7 gives the relative abundance of several elements as a function of the current density of baryons – a type of particle that absorbs and emits light (like protons and neutrons) – in the universe. As can be seen from these results, if we use the D/H value to approximate the density of the universe, there is not enough material to stop the expansion of the universe. The density estimated from the D/H value is some 10–30 times too small. (The detailed value depends somewhat on the value of our old friend H_0, but the conclusions remain the same.)

This startling result leaves us with two broad possibilities. The first is the straightforward interpretation that the universe had a beginning with the Big Bang, but will simply keep expanding forever. Thus, it would have no ending, no 'Big Crunch'. This result is scientifically acceptable, but many people are uncomfortable with it on philosophical grounds.

Figure 5.7. Abundances of elements – He (helium), D (deuterium), and Li (lithium) – as calculated from a standard Big Bang model. The abundances are plotted against the density of baryons in the universe in units of the closure density (Ω_0). The observed value of D/H is plotted, as is the estimated range for primordial D/H. The highest baryon density possible is about one-tenth of the closure density. *(J. Linsky, Joint Institute for Laboratory Astrophysics and University of Colorado)*

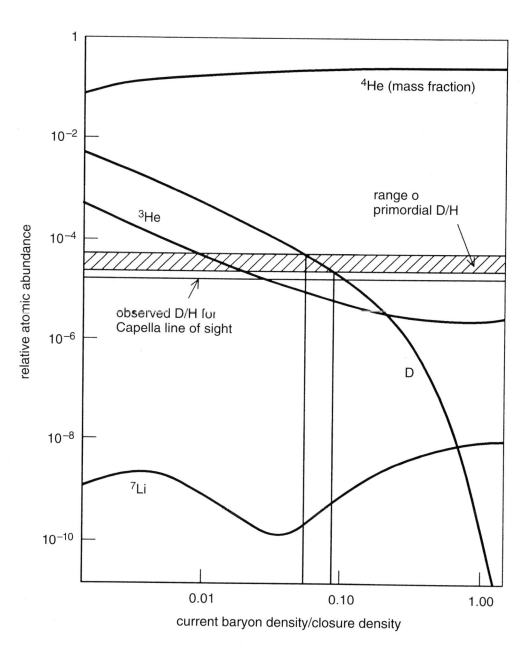

The second possibility is that the universe will stop expanding, and that ours is a closed universe. If that is true, then things become truly bizarre. So far, we have been talking about ordinary matter, composed of baryons; this so-called ordinary matter interacts with light. But is there matter that does not interact with light? This brings us back to the question of the dark matter.

Astronomers know from the motions of galaxies in clusters, and from mass estimates of galaxies based on rotation curves, that there is a lot of unseen matter in the universe. So, could the rest of the mass of the universe be in the form of exotic, non-baryonic material that does not emit radiation at any wavelength? As we discussed earlier, candidates for such matter include heavy neutrinos, axions, and WIMPS. The particle physics community is actively searching for such forms of matter; however, such matter has not yet been detected by nuclear physics experiments.

If the universe is closed, then the ordinary matter that we are familiar with can only comprise 2–10 per cent of the universal total. The rest, between 90 and 98 per cent, would be in an exotic form that is presently unknown. If this scenario is the correct one, it would force the liberation of the human species from its ancient fascination with the geocentric hypothesis and expose our own baryonic chauvinism. Not only are we not the center of the universe, but we may not even be made of the dominant form of material in the universe!

Until such time as we understand the more exotic aspects of baryonic versus non-baryonic matter, we still have much work to do with the observations we have made on matter we can detect. If the D/H measurement and the inferred value of Ω_0 stand the test of time, they will have important cosmological consequences. In addition, the D/H measurement is a prime example of a very local measurement with cosmological implications.

Determination of distances

HST's contribution to solving distance determinations in cosmology comes about in two ways: by measuring stellar parallaxes of stars relatively near the Sun, and by using standard candles to measure very great distances. In these two ways, astronomers take steps toward establishing a more accurate extragalactic distance scale. The Astrometry Team (using the Fine Guidance Sensors) is making very accurate parallax measurements. However, due to the inherent length of time required for such accuracy, they are concentrating on specific parallaxes, as opposed to the *Hipparcos* satellite, which did an automated survey. Still, HST parallaxes with accuracies of 0.002 arcsecond are possible from HST, whereas ground-based measurement of parallaxes take decades, and are less accurate.

The importance of these parallaxes for stars relatively near the Sun to the extragalactic distance scale is as follows. The scale is built up like a ladder one rung at a time, or, in the case of Freedman's Eiffel Tower, one level at a time. Distances to nearby stars are used to determine the measuring techniques for stars farther away. Each time distances are established, they are used to calibrate the technique for objects farther away. But, each step rests on the one below it, and an error in the distances to stars near the Sun could throw off the whole progression of distance determinations. Thus, while the astrometric work being carried out with the FGSs may not seem spectacular, it is very important.

The study of distant galaxies with the goal of establishing the value of H_0 to within 10 per cent is one of the so-called key projects given high priority for work on HST. Ultimately, HST will study many galaxies with a variety of techniques. These distances, along with a series of redshift measurements, will allow astronomers to shore up one of the shakier steps on the cosmological distance ladder, and, finally, to determine a secure value for the Hubble Constant. We may be able to declare victory when there is a general consensus that the different approaches have converged on a small range of values. This is not currently the case, but progress is begin made.

The HST work on H_0 effectively began in 1994 when astronomers announced measurements of distances to Cepheid variables in the Virgo Cluster galaxy M100 (Figure 5.8). The galaxy was determined to be some 56 million light years away instead of 50 million light years as previously thought. On the basis of this distance, a new value for H_0 of 80 kilometers per second per megaparsec was calculated. This number immediately produced controversy because it was higher than some astronomers expected. It led to a low range of values for the age of the universe: 12 billion years for a low-mass universe expanding constantly since the Big Bang, or 8 billion years for a universe containing the critical density of $\Omega_0 = 1$, which expanded faster during the earlier stages of the universe. Clearly, something seems amiss. As more than one scientist has said, it is a wonderful time to be a cosmologist.

Of course, we should not base too much on a distance to a single galaxy, and so HST astronomers are well on their way to measuring distances to many galaxies. Figure 5.9 shows the WF/PC-2 image of NGC 4639, another galaxy in the Virgo Cluster. The Cepheids in this galaxy were used to determine an accurate distance. This study is important for two reasons. First, the Virgo Cluster is large, and its 'distance' depends somewhat on estimates for the distances of individual galaxies from its center. Directly determined distances for a number of Virgo Cluster members will eventually remove this concern by giving us solid data to consider. Secondly, (and fortuitously), NGC 4639 has a Type-Ia supernova (SN 1990 N) which can be

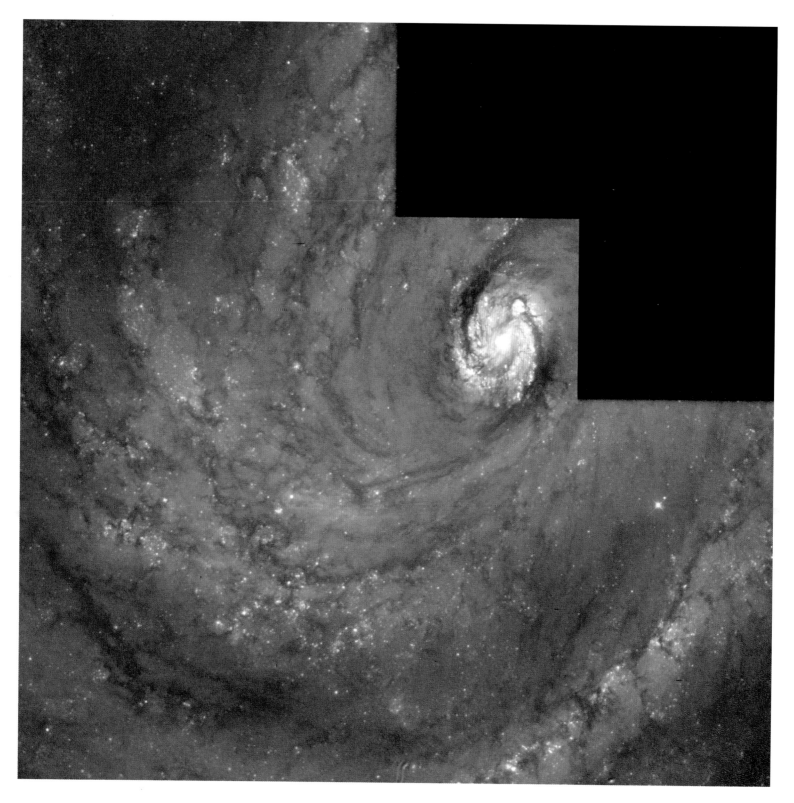

Figure 5.8. WF/PC-2 color composite image of
M100 showing the resolution of individual stars
in the spiral arms. *(J. Trauger, Jet Propulsion
Laboratory; NASA)*

Figure 5.9. WF/PC-2 image of the galaxy NGC 4639 in the Virgo Cluster. The blue 'dots' in the outlying regions indicate young stars, including Cepheids. This galaxy also has a supernova of the type used as a distance indicator. *(A. Sandage, Carnegie Observatories; A.Saha, STScI; G.A. Tanman, L. Labhardt, Astronomical Institute, University of Basel); F.D. Macchetto, N. Panagia, A.Saha, STScI; ESA, NASA)*

Figure 5.10. WF/PC-2 image **(above)** of the barred spiral galaxy NGC 1365 in the Fornax cluster of galaxies. **(Opposite)** The instrument focused on a region in the lower spiral arm populated by young stars. This region is a likely place to search for Cepheids. About 50 have been found.

Table 5.2 *Values of H_0*

80±17 (W.L. Freedman)
78±11 (W.B. Sparks, *et al.*)
55±10 (G.A. Tammann, *et al.*)
70±9 (S. Sakai, *et al.*)
69±8 (N.R. Tanvir)

Note: H_0 is in units of kilometers per second per megaparsec. Data based on papers collected in the proceedings of the Second Science With the Hubble Space Telescope meeting, held in Paris, France, December 1995.

used as a distance indicator. Thus, a nice side-effect is that the accurate distance from Cepheids calibrates the Type-Ia supernova for use in determining distances to other galaxies much farther away than NGC 4639.

Figure 5.10 presents an image of the barred spiral galaxy NGC 1365, which is located in the Fornax cluster of galaxies. Approximately 50 Cepheids have been discovered, and their distances have been determined to be 60 million light years away. This result is important for two reasons. First, the Fornax Cluster is more compact than the Virgo Cluster; this means a smaller range of uncertainty for the distances of member galaxies from the cluster's center. Secondly, the Fornax and Virgo Clusters are in almost opposite directions on the sky, and systematic errors associated with making measurements in a particular celestial direction can be eliminated.

So, what then is the situation for the value of H_0? We may approach this question by thinking of it as a 'good news/bad news' scenario.

First, the good news. All of the values in Table 5.2 are remarkably close together by historical standards (discussed below). The difference between the extreme values (80−55) divided by their average [(80+55)/2] gives a difference of 37 per cent. If this expressed as an error bar, we would have 67.5 ± 12.5, or a margin of error of 18.5 per cent.

This number is the latest in a string of estimates that began with Edwin Hubble. In 1936, in 'Realm of the Nebulae" he proposed a value for H_0 of 530 kilometers per second per megaparsec. This value caused great problems because it implied that the universe was only about 2 billion years old, younger than the age of terrestrial rocks. The techniques have been refined over the decades, errors have been eliminated, and now we have a historically comfortable situation. If, for purposes of illustration, we adopt $H_0 = 67.5$ kilometers per second per megaparsec, the maximum age of the universe (under most assumptions) is 14.8 billion years, and

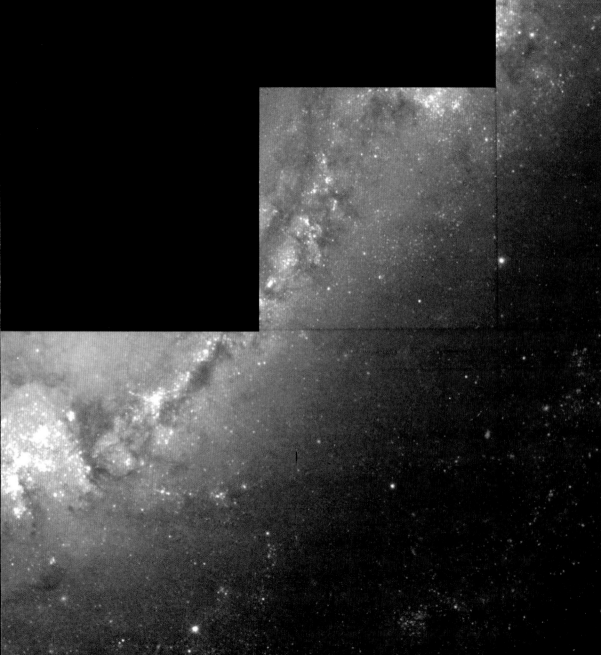

the age for $\Omega_0 = 1$ is 9.9 billion years. The problem with the ages is greatly moderated, particularly when we consider that most evidence points toward a low-density universe. Thus, the larger value of 14.8 billion years should be used for comparisons.

So what is the bad news? Determinations of the value of H_0 have developed such that there are those who favor the higher values, 69 to 80 kilometers per second per megaparsec, and those who support the lower value, 55 kilometers per second per megaparsec. Advocates of the different views vigorously defend their territory, but, as with most such scientific disagreements, the issue will be resolved with the passage of time and the acquisition of new data. The Key Project of HST for H_0 is to determine accurate distances to about 30 galaxies; more than half of these observations have already been completed. In a few more years, we will see the completion of the Key Project Survey and convergence on the value of H_0. We will know that we are close when the H_0 values from Cepheid variables, Type-Ia supernovae, other secondary indicators, and timing of quasar light variations in gravitationally lensed situations (discussed below) consistently fall into a small range.

Distant galaxies and clusters

From time to time, we see an announcement of the discovery of the most distant galaxy known or a new observation of a galaxy in one of the most distant cluster of galaxies. The Hubble Deep Field, discussed in Chapter 4, opens the door to new discoveries in this area.

Figure 5.11 shows a very distant galaxy in the Hubble Deep Field. Among the nearly 3000 galaxies in the images, several dozen galaxies appear to be more distant than the farthest quasars, which are usually the record holders for the farthest distance. The distance estimates are based on the colors of the galaxies, using the same principle as the 'ultraviolet dropout' effect described in Chapter 4 (see Figure 4.13, page 142).

The galaxy indicated in Figure 5.11 existed when the universe was less than 5 per cent of its current age and could have formed only a few hundred million years after the Big Bang. The

Figure 5.11. A small section of the Hubble Deep Field shows a very faint galaxy (arrowed) that appears to be the most distant ever discovered. *(K. Lanzetta, State University of New York Stony Brook; NASA)*

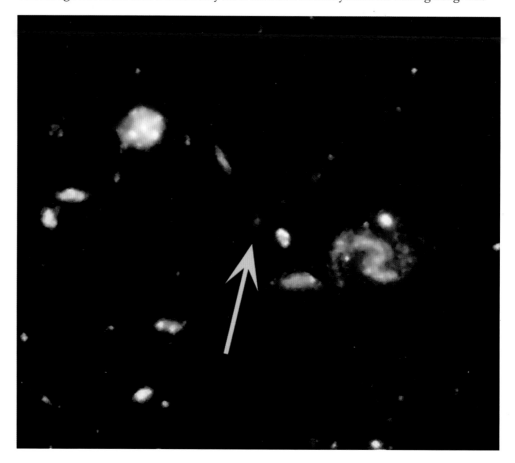

detection of these galaxies also begins to indicate how many were forming very early in the universe.

Recall that since light travels at a finite speed, these galaxies are seen as they were in the distant past. Thus, we are seeing the birth and growth of galaxies over cosmic time. These results have implications bearing not only on the formation and evolution of galaxies but also on the ultimate fate of the universe.

Analysis of these most distant galaxies for cosmological purposes is the best hope for directly determining the deceleration parameter, Q_0. If galaxies at high redshifts can be resolved, their supernovae can be observed, and then distances can be determined. If, however, the density of the universe is low (corresponding to a value for Ω_0 much lower than unity), the deceleration of the universe may also be very low. The obvious and serious complicating factor is that stellar evolution may be considerably different in very early galaxies, and, hence, our distance indicators may be inaccurate.

Another promising approach to studying very distant galaxies takes advantage of the gravitational lens effect described in the following section. Massive objects, such as clusters of galaxies, warp the space around them and focus the light from even more distant objects. A good example of this type of object shows up in the WF/PC-2 image in Figure 5.12. The brighter objects in the image are galaxies in a foreground cluster that is 'only' 5 billion light years away. The lensed, more distant galaxy is the orange/red arc that appears to the lower right of center in the full view (below, left); a close-up view is shown at upper right. Spectroscopic observations with one of the Keck 10-meter telescopes on Mauna Kea, Hawaii, established the great distance of this galaxy. The high redshift, $z = 4.92$, means that the galaxy is about 13 billion light years away, or that the galaxy was shining only a billion years or so after the Big Bang with a brillance of more than ten times that of our Milky Way Galaxy.

The lensed image shows details some five to ten times finer than could be obtained by HST alone. Very compact regions of intense star formation – some 700 light years across – are clearly seen in the original image (below) and in the 'corrected' image (lower right). In the

Figure 5.12. The gravitationally lensed images of a very high redshift galaxy. This WF/PC-2 view shows the galaxy in the full image **(left)**, in a close-up view **(upper right)**, and in a close-up view **(lower right)** that attempts to give the 'normal' appearance by using a model to remove distortion produced by the lensing. See text for discussion. *(M. Franx, Kapteyn Astronomical Institute; G. Illingworth, Lick Observatory; NASA)*

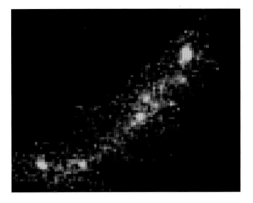

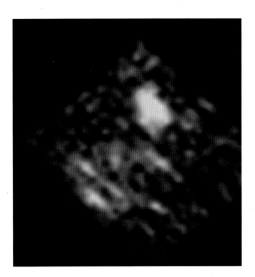

latter image, an attempt was made to remove the distortion produced by the gravitational lensing and to show how this very young galaxy would normally appear. The images show star formation occurring at a much faster rate than in most galaxies at the current epoch.

The need to go to longer and longer wavelengths because of the larger and larger redshifts of early, distant objects emphasizes the importance of expanding our observational capability into the infrared. The installation of the Near-Infrared Camera and Multi-Object Spectometer into HST in February 1997 gave us this ability by extending wavelength coverage to 2.5 microns (25 000 angstroms). In the very near future, infrared Hubble Deep Field images will provide a wonderful source of data for studying very young galaxies. If the Next Generation Space Telescope (NGST; see Chapter 6) is launched in the next century, the coverage will be extended to at least 5 microns (50 000 angstroms !), giving us an even better window onto the very early universe.

General relativity and gravitational lenses

The Theory of General Relativity is the basis for calculations of the evolution of a Big Bang universe. Confirmation of the general relativistic picture is always welcome. According to general relativity, the space (strictly speaking, the space-time) near a massive object is distorted by the object. This produces the same basic effect as a gravitational pull, and orbits can be produced. In fact, for planets, general relativity and Newtonian mechanics give essentially the same results.

This distortion of space has another consequence. The light shining from an object directly behind another, more massive, object can have its direction changed so that the object itself appears to the observer to be shifted from its real position. This is a phenomenon called a 'gravitational lens effect'. HST is carrying out a number of observations of these lenses, and

Figure 5.13. Schematic diagram illustrating the gravitational lens effect. Light from a distant galaxy can appear as multiple images due to the lensing effect of foreground galaxies. *(Richard Ellis, University of Cambridge; NASA)*

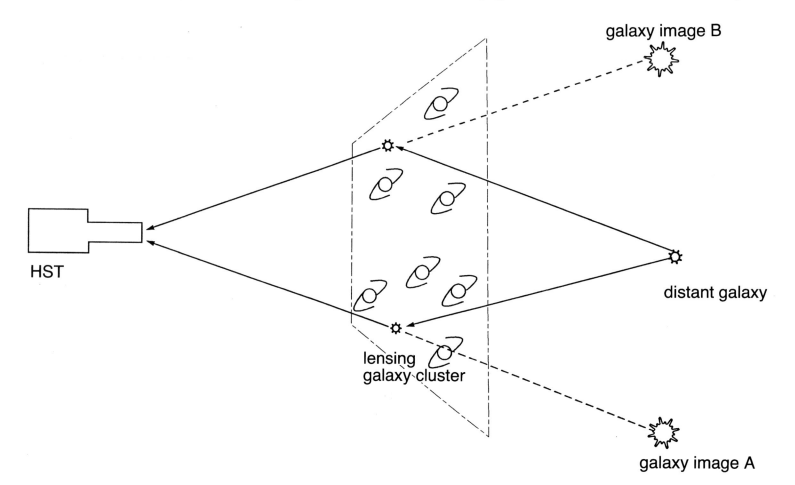

one on-going program is being led by John Bahcall. For purposes of his study, HST images provide Bahcall with unprecedented information about the frequency of gravitational lensing by galaxies. This program was dubbed the 'Snapshot Survey' because it uses HST's Planetary Camera to take short exposures during gaps in the telescope's schedule. The resulting images are slightly trailed, but, according to Bahcall, the data are quite valuable for determining characteristics of lensing objects: 'More HST data has been studied in this survey than any other HST science program.'

HST's other gravitational lens programs are returning spectacular images. Figure 5.13 shows the situation schematically for the case of a very distant galaxy seen through a cluster of foreground galaxies which act as the lens. Several images can result from a lensing situation, depending on the precise details of the alignment. Figure 5.14 shows the HST view of the rich galaxy cluster Abell 2218. In this gravitational lens, the telescope's high resolution revealed a collection of ghostly looking images. These arcs were created by the distortion of light by the galaxy cluster. The cluster is so massive and compact that light rays from a very distant population of galaxies are bent to form the arcs of light shining around several of the cluster members. This distant set of galaxies, which lies directly behind the lensing cluster, existed when the universe was just one-quarter of its present age. Since we see these galaxies at a very young age, the arcs can be used to look at star-forming regions in the distant galaxies. Besides studying the larger arcs, astronomers are also interested in a collection of 'arclets' that also appear in the image. Distances will be calculated for 120 of the arclets, which represent galaxies 50 times fainter than can be seen using ground-based instruments.

Another gravitational lens is shown for the cluster 0024+1654 in Figure 5.15. The blue, loop-shaped objects are multiple images of the same primeval galaxy produced by the yellow elliptical and spiral galaxies in the cluster near the center of the photograph. The cluster is approximately 5 billion light years away in the constellation Pisces, and the blue galaxy is at

Figure 5.14. WF/PC-2 image of the rich galaxy cluster Abell 2218, which contains several examples of gravitational lensing. The arc-like pattern spread across the image like a spider's web is an illusion caused by the gravitational field of the cluster. *(W. Couch, University of New South Wales; R. Ellis, University of Cambridge; NASA)*

about twice this distance . The bits of white imbedded in the blue galaxy are thought to be young stars, and the dark core inside the rings is thought to be dust from which stars are formed. This information, the blue color and the lumpy appearance, suggests that this is a young galaxy where star formation is quite active.

The discovery of different gravitational lenses is exciting, and studying them offers the potential for very important cosmological results. A light variation (for example, the explosion of a supernova) in the background galaxy, or a change in quasar brightness, will be seen at Earth at different times for different gravitational-lens-produced images. If the geometrical circumstances are known, the time delay gives an independent value for the Hubble Constant, H_0. This type of investigation requires careful study of the source for the light variations, and we await convincing results, which may come from ground-based observations.

Where are we now?

We opined earlier that cosmology and general relativity are two of the most profound types of science engendered by a study of astronomy. It is important to remember that cosmology was one of the main drivers for HST back when the telescope was still a twinkle in astronomer Lyman Spitzer's eye. Scientifically, we end up back at the beginning of HST's life story. If HST did not do anything else except help solve the cosmological distance problem, it would still be counted as one of the great successes of astronomy.

As we have seen throughout this chapter, HST results have spiced up the stew that is our understanding of the universe. The refined value for the deuterium to hydrogen ratio, which limits the value of Ω_0 for baryons in the universe to approximately 0.05, is one result with interesting implications. The value of the Hubble constant, H_0, of about 70 kilometers per second per megaparsec, implies an age for the universe of 9.5 to 14.3 billion years. The age depends on a more accurate value of Ω_0 for all matter (the value of H_0 does not depend on Ω_0, but T_0 does).

A reasonable, but by no means definitive or final, summary of the current situation is that we are in a low-density universe with a Ω_0 value of about 0.25. The low density implies that the universe will expand forever and that the deceleration parameter (Q_0) is small. If true, the Hubble expansion has not been significantly slowed since the Big Bang, and hence a Hubble Constant of 70 kilometers per second per megaparsec yields an age for the universe of about 14 billion years. If we determine a Hubble Constant of 70 kilometers per second per megaparsec and a low-density universe, there is no conflict between an age for the universe of 14.3 billion years and the potentially troublesome problem of 13 to 15 billion year old stars in globular clusters.

However, if $\Omega_0 = 1$, a 9.5 billion year old universe might represent a conflict with the ages of globular cluster stars, even though there could easily be some uncertainty in the star ages. This possible conflict in ages has led some scientists to resurrect the 'Cosmological Constant', a concept introduced many years ago by Albert Einstein, representing an acceleration of the Hubble expansion. If this idea is correct, the universe is now expanding at its *fastest* rate, the average rate of expansion is *lower* than measured by the (current) H_0, and the universe can be *older* than implied by H_0. This is contrary to the case where there is enough matter in the universe to slow the expansion ($\Omega_0 = 1$), and hence the universe is now expanding at its *slowest* rate, the average rate of expansion is *higher* than measured by the (current) H_0, and the universe can be *younger* than implied by H_0.

All of this discussion emphasizes the critical importance of determining the amount and nature of the dark matter in the universe, and determining a reliable value for Ω_0. In this investigation, the HST community may need major assistance from the particle astrophysics community. An abundance of exotic particles would need to exist to bring the value of Ω_0 up to unity (the critical value needed to slow the expansion of the universe), but even value of 0.25 requires a major contribution from exotic particles.

Figure 5.15. **(Opposite)** A marvellous WF/PC-2 image of a gravitational lens in the galaxy cluster 0024+1654. The blue, loop-shaped objects are lensed copies of a distant galaxy, behind the foreground cluster. *(W. Colley, E. Turner, Princeton University; J.A. Tyson, AT&T Bell Labs; NASA)*

Another unresolved problem is how the universe formed structures such as galaxies, clusters of galaxies, etc. in the times available and with the density fluctuations expected. Until all the pieces of this puzzle fit together, our picture is not final.

For now, until the HST can uncover convincing evidence to the contrary, we are constrained to live in a universe that is expanding forever. It may be a younger universe than we thought, and some of our ideas about the ages of the oldest globular cluster stars need to be re-examined. Even our studies of the Sun, a star we know better than any other in the universe, reflect a great lack of knowledge. Based on our theories of stellar evolution, we should be seeing many more neutrinos from the Sun – yet we see so few (about one-third the number expected) that we are questioning our concepts about the Sun's energy source and our knowledge of elementary particle physics. If we can be this imprecise about the Sun, then a possible few-billion-year discrepancy in the ages of certain stars in the universe is understandable. What it points out is that we do not know everything.

Not only is the universe that HST shows us not always cooperating with the theories we have developed to explain it – it is even stranger than we ever dared imagine. If HST never accomplishes anything more, this will be its legacy – and something astronomers everywhere can look to as a source of extraordinary, exciting, and dynamic discoveries.

The once and future Space Telescope

In 1923, German rocket scientist Hermann Oberth outlined ideas for a space station and an orbital space telescope in his book *Die Rakete zu den Planeträumen* (*The Rocket Into Planetary Space*). He imagined a telescope attached to a station in geosynchronous orbit (where today's telecommunications and weather satellites are located), and even discussed the effect of station jitter on observations of dim, distant objects. To solve that problem, he came up with some rather ingenious ideas for attaching a space-based telescope to a small asteroid, and he even suggested that future scientists tow the asteroid to Earth orbit so that the telescope 'night crew' would not be bored while living on a station orbiting between Mars and Jupiter!

Oberth's writings attracted a great deal of attention and criticism at a time when any idea of traveling beyond the atmosphere was best left to people who were more familiar with

There is a solution to every problem that is simple, elegant and wrong.

attributed to H.L. Mencken

We are all in the gutter, but some of us are looking at the stars.

Oscar Wilde

Figure 6.1. Hermann Oberth (1894–1989) and his daughter viewing an exhibit on the Space Telescope at the National Air and Space Museum (Washington, D.C.). *(Courtesy of the National Air and Space Museum, Smithsonian Institution)*

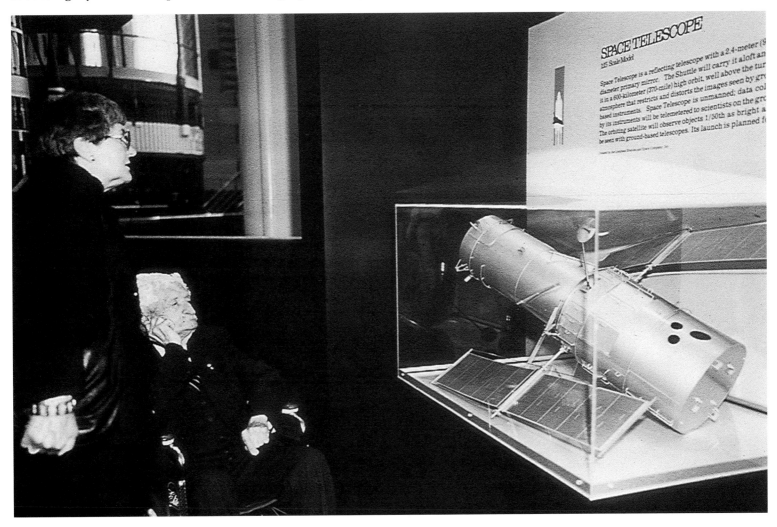

such crazy ideas – science-fiction writers. However, he persisted in his thinking, and revisited the idea of a large space telescope in 1957, in a book called *Menschen Im Welträum* (*Man In Space*), writing

> *It is only too obvious that astronomical research would gain remarkable benefits from a space station. The atmosphere, gravitation, and the movement and rotation of the Earth affect observation of the sky. The immense difficulties involved in producing even larger telescopes, and the fact that their use may be interfered with by the atmosphere have been shown quite recently in the case of the 200-inch Mount Palomar telescope.*

Figure 6.2. The HST primary mirror at Perkin-Elmer (now Hughes-Danbury). *(STScI)*

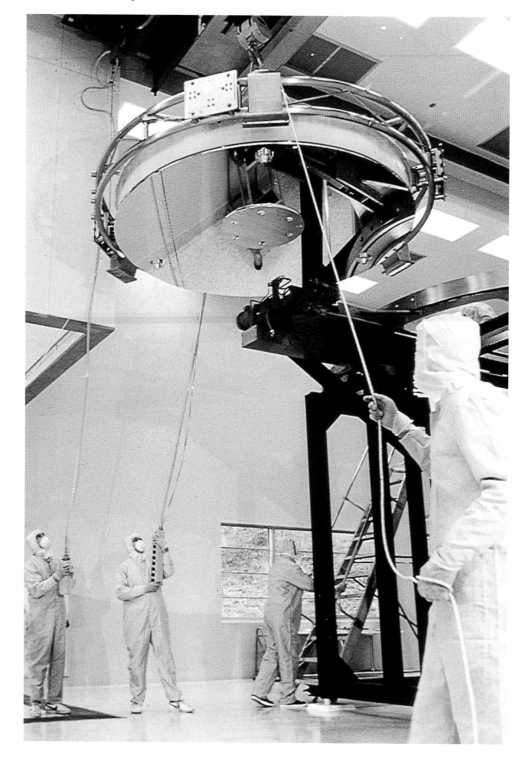

Oberth's words seem prophetic to us now, because 40 years after he wrote them, a large orbiting observatory *is* benefitting astronomical research. As we have seen in the preceding chapters, HST has made incredible contributions to nearly every area of astronomical endeavor, and there is every indication that it will continue to do so, perhaps well past its 15-year planned lifetime. Whether HST shuts down or not, its work will be extended by a telescope that looks very much like Oberth's original design – a large, floating mirror in space. It will be the Next Generation Space Telescope, and it will pick up HST's charge to investigate the universe out to the limits of our observational capability.

First light, first frustrations

Imagine that somewhere in the world, probably atop a mountain, well away from light pollution, a group of astronomers, engineers, night assistants, and electronics experts angle a telescope toward the sky for the first time. If everything has been built to specification – the mirror is ground correctly, the instruments are aligned precisely, and the pointing programs execute accurately – the telescope will open its eye to the universe. The coordinates of some distant planet, or star, or galaxy have been programmed into the telescope's drive computer.

When all is ready, the observation begins. Light from that distant object enters the telescope tube, bounces off the mirror and into the apertures of complex instruments designed to analyze and transform it into data. Years of planning, building, tweaking, experimenting with new technology, all boils down to the moment of truth, the event called 'first light'.

Generally, first light for any kind of telescope is a relatively private moment. Only the astronomers, night assistants, and designers are there to see the first images from a newly commissioned facility. Actually, first light is most important to the people who will spend their time running the telescope, learning its nuances, and correcting its glitches. No members of the press are waiting outside the door, breathless for pictures. There are no government dignitaries, no administrators, just the people who designed the telescope and its instruments. First light for a telescope is like first flight for a new aircraft, or a ship's shakedown cruise. For the telescope's users, first light is an astronomer's version of the big dress rehearsal for all the nights to come.

If the first-light images from a ground-based telescope are not the crisp, crystal-clear pictures everyone is used to seeing in the pages of astronomy magazines, it is not exactly a surprise. First-light images rarely look spectacular, but they *are* important because they are really test images that tell astronomers how the telescope is performing.

May 20, 1990, was the Hubble Space Telescope's 'first light'. The event was like nothing any other telescope team had ever experienced. It was an exhilarating and confusing time, and the culmination of years of planning, development, and training by thousands of people.

At the appointed time, HST – from its orbital vantage point 380 nautical miles above the Earth's surface – turned its flywheels and gracefully moved into position to catch the light from a star cluster over 1300 light years away. Deep inside the telescope, relays clicked, motors whirred, electronic circuits came to life, and HST signaled its readiness. In control rooms at the Goddard Space Flight Center in Greenbelt, Maryland, the Marshall Space Flight Center in Huntsville, Alabama, and the Space Telescope Science Institute in Baltimore, Maryland, everyone watched computer screens as the pre-programmed test observation sequence executed. At Goddard, members of the press watched the scientists, ready to translate the experience for the waiting public.

It was a moment scientists and technical teams had been anticipating for years. To be sure, no one expected the first-light images to be very exciting or spectacular. They were more for the benefit of the engineers and controllers to evaluate the telescope's performance.

The HST teams were not the only ones excited about the event. Fueled by a never-ending stream of 'Hubble Hype' in the press, the general public was ready to be amazed by glorious astronomical images. For months, HST was subjected to the most impressive amount of publicity since the first Moon landings. It was going to be *the* telescope to reveal the secrets of the stars, probe the innards of galaxies, and give scientists a wide-ranging eye on the universe.

Far above Earth, light from the star cluster NGC 3532, located in the southern hemisphere constellation Carina, streamed into the telescope. It hit the primary mirror, and bounced onto a smaller secondary mirror. That image of NGC 3532 was hailed as a great success by the press. The image was not the prettiest sight in the history of astronomy, and everyone knew that the telescope needed some focusing adjustments, but the first step had been taken.

Star images normally look like a bright spot of light surrounded by a little scattered light. This is because most of the star's light is focused by the telescope's mirrors into a little 'core' or bright spot. HST's first-light image showed a sharp spike of light surrounded by a huge halo and strange tendrils of light extending out from the central core – *not* the way a star should look through a correctly focused telescope. As the telescope relayed more images to the ground for analysis, everyone's uneasiness increased.

A crescendo of concern first began to build among HST team members after a project meeting held at Goddard Space Flight Center on the day after first light. During that meeting, uncertainties about the mirror were aired in a discussion about the way the images looked. Team member Roger Lynds – an astronomer and optical expert – stood up and said he thought HST had a serious spherical aberration problem. It was the first time anyone mentioned those words in connection with HST's main mirror, and Lynds's suggestion was dismissed abruptly. Yet, Space Telescope Science Institute scientist Chris Burrows came to the same conclusion almost immediately when he started to analyze the first-light image and realized that something was curiously wrong with it. He wrote a software program that calculated the kinds of problems that would create such an image. The report based on his analysis independently confirmed Lynds's diagnosis of spherical aberration.

Of course, no one wanted to hear such a depressing diagnosis because it implied that a major mistake had been made. A spherically aberrated mirror is one that has been ground incorrectly. The curve at the center of the mirror does not match the curve at the edge. Essentially, the flawed mirror does not reflect light into a tiny point of encircled energy called the 'core'. Instead, it spreads the light out over a larger area. On HST, the center focused well, but the edge was out of focus, making a big halo.

It was simply unthinkable that HST's main mirror, the 'heart' of the telescope, could have been ground incorrectly without someone noticing before launch. No one – least of all Hughes-Danbury Optical Systems (formerly Perkin-Elmer), the contractors responsible for the mirror, wanted to admit to that kind of mistake. Throughout May and June, a series of tests were run to diagnose the mirror problems. On June 19, the inevitable conclusion was arrived at: spherical aberration.

At that point, it was all over except the shouting. The word spread through the teams like wildfire, but, to everyone's great amazement, the news had not reached the press. Certainly there had been comments circulating about 'tendrils' in the images – hints that some science writers have said in retrospect should have tipped them off to a serious focusing problem. Aside from the scientists on the scene who were working with the engineers from Hughes-Danbury to understand just what they were seeing, very few people in the larger astronomical community understood just how serious the problem was. The actual announcement was left to NASA management, and a press conference was arranged for June 27.

The press conference was a funereal and depressing event, with NASA scientists almost at a loss for words to describe their feelings. Anyone watching the conference was left with the immediate feeling that HST was dead in the water. The press picked up on that mood immediately, setting the stage for highly critical hyperbole.

Taking advantage of the media spotlight, politicians and pundits rushed to denounce the project as a waste of money, an outrage, and a 'techno-turkey'. Almost overnight, political cartoonists and late-night talk show hosts were making jokes at the telescope's expense. The long-awaited telescope that would see to the distant reaches of the universe became linked in the public's mind with bungling, government-sponsored science, a sort of orbital Mr Magoo.

The aberration was indeed an embarrassment of the worst sort. It was the equivalent of a rookie mistake, caused by a mix of technical problems, politics, and just plain human stubbornness in the face of error. It should not have happened, but it did.

Every scientist with time on the telescope had questions about how the aberration would affect their observations, the cameras, and the spectrographs. Everyone expected the worst, and it looked like astronomers were left standing at the door to the universe, peering at the stars with myopic eyes while waiting for beleaguered optical experts and engineers to solve a completely unexpected mystery – 'The Case of the Misground Mirror'.

Those were depressing days for the people connected with Hubble Space Telescope. Everyone involved with the project was in a state of grief and disappointment over what looked like a colossal missed opportunity. Ironically, there occurred a loss of focus, not just for the mirror, but in the observing programs of scientists who had awaited HST for years. There was lost science and a flagging confidence in that 'can-do' attitude that had been the hallmark of the American space effort.

Most embarrassing of all was the inexorable erosion of public interest in what astronomers do best – unlocking the secrets of the universe and showing them off to an interested community. With the staggering blow of a spherically aberrated, jittery, cranky telescope, dreams of exploring the ends of the universe seemed to be turning into just so many ashes on the dustheap of failed science. The promised wonders of the universe that HST's designers and promoters had predicted were starting to look like pipe dreams.

The big question was: what do we do now? Incredibly, despite the spherical aberration and other problems with the telescope, scientists found that the telescope was not a total loss. It could still do observations, but it focused only 10–15 per cent of the light into a rather sloppy central core, instead of the tightly focused 75 or 80 per cent expected by astronomers. The rest was scattered into tendrils and spikes. If something could be salvaged from the data that HST could provide, how would astronomers go about digging it out?

Science and the human spirit have always flourished through adversity, and – as we all know – HST was not a lost cause. Quietly, behind the scenes and largely out of the media spotlight, some of the brightest people in the science business turned their attention to figuring out ways to compensate for the telescope's flaws, even before the big news of HST's problems was released. One possibility, suggested by Kitt Peak National Observatory scientist Tod Lauer, was to use 'deconvolution', a computerized image processing technique that essentially puts the light back where it belongs. The drawback was that if the data were deconvolved too much, useful information might be lost in the process. However, deconvolution could turn lousy images into acceptable ones. If examples of deconvolved images had been available on the day spherical aberration was announced, it is possible that some of the publicity HST endured would not have been quite so unsavory. Another particularly useful approach was to increase exposure times, gathering more light from an object to increase the data rate.

These processes later became important parts of the astronomer's toolbox during HST's first years on orbit, but, in the dark days of summer 1990, few people outside of the HST community knew about them. Instead, everyone wanted to know, How did this happen? How could the main mirror of a such an expensive telescope be ground wrong? Was not somebody paying attention?

Grinding a telescope mirror is an arcane art, as many amateur astronomers find out when they get the urge to make their own telescopes. The optical engineer starts out with a 'blank' – a piece of optical-quality glass. That glass has to be ground and polished repeatedly until it is the proper shape to focus light. When all the grinding and polishing is complete, a reflective coating is applied. If this is done correctly, the primary mirror will focus most incoming light into a tiny point. This well-focused core of light is the 'Holy Grail' for the scientists using the telescope. The better the focus, the better the information.

HST's main mirror was ground by a team of optical engineers at Perkin-Elmer Corporation of Danbury, Connecticut. To make sure that the mirror was ground and polished accurately, Perkin-Elmer built a special device called a 'reflective null corrector'. This consisted of two

Figure 6.3. Kitt Peak National Observatory near Tucson, Arizona. *(National Optical Astronomy Observatories)*

small mirrors and a tiny lens hung above HST's main mirror. A light beam was flashed onto the primary mirror through the tiny lens in the corrector. This test created a pattern of light called an interferogram which looked like a big fingerprint. If the 'fingerprint' did not look right, opticians would continue to grind the mirror until it looked as expected.

This process should have guaranteed a perfect mirror, but, in installing the lens in the null corrector, the technicians made a mistake. The lens had to be set at a precise distance from the main mirror. To determine that distance, the technicians used a rod with a special cap that had a tiny hole in it. An alignment light beam was sent through the hole in the cap down to the lens. To make sure that the laser beam did not bounce off the cap, special paint was applied to it. The light was supposed to go through the hole and bounce off the tip of a measuring rod, which would then tell the technicians where to set the lens. If it failed to do this, they would know right away to re-aim the light.

It was an intricate measuring process, and it failed because a tiny spot of paint had worn off the cap. When the light went down to the cap, it bounced off the worn spot instead of going through the hole down to the tip of the rod. To the technicians, it looked like the light was bouncing off the rod and so they attempted to hand-correct the placement of the lens. When the lens would not go where the corrector said it should, the technicians inserted three little washers into the setup to make it work. The difference in distance between where the lens was set and where it should have been set was only 1.3 millimeters, but it was enough to throw off the whole process. The result was that the opticians thought they needed to grind off more glass. It was a near-fatal mistake. They made a perfectly ground mirror – but it was ground perfectly wrong. As one scientist put it, 'Hubble Space Telescope had the best spherical aberration money could buy'.

Fortunately, as we shall see later in this chapter, the basis for a correction of the spherical aberration lay in the very optics which stymied the mirror in the first place. In fact, it was a fix very much like putting eyeglasses on a near-sighted patient. In this case, however, the 'patient' was on-orbit above the Earth, and the 'eye' needing correction was quite large.

Misgrinding, then, was the technical reason why HST has an aberrated mirror. However, people, not computers, did that technical work, and understanding the human factor in Hubble's troubles is important. Simply put, HST faced technical hurdles and entrenched managerial problems even before it was designed. Every complex human institution has such problems – but they are particularly hard to accept with a telescope that started out with such high expectations. HST did not start out as a jittery, spherically aberrated mess. In fact, it started out as a solution to one of the problems astronomers deal with: the atmosphere.

HST's place in history

Until the space age, astronomers plied their trade from the surface of the Earth. They worked behind the large, unblinking eyes of US-based facilities like Arizona's Kitt Peak National Observatory, the observatories atop Mauna Kea in Hawaii, California's Lick Observatory, Mount Palomar, and Mount Wilson, Wisconsin's Yerkes Observatory, and a variety of excellent institutions around the world: the Anglo-Australian Observatory, the Royal Observatory at Edinburgh, European Southern Observatory in Chile, Armagh Observatory in Northern Ireland, and the international facilities at Cerro Tololo Inter-American Observatory.

Magnificent science has been, and continues to be, done at these facilities, which are the backbone of astronomy research. In fact, the astronomers who use HST also use ground-based facilities, and are well-versed in achieving 'good science' at those observatories.

Even the best ground-based facility, however, is hampered by the blanket of gases that we breathe to keep alive – the Earth's atmosphere. To put it simply, for some types of astronomy, the atmosphere is a big problem. It interferes with the light that astronomers want to see in three distinct ways: it bends it, it absorbs certain wavelengths, and it also gives off radiation that interferes with incoming light.

Scientists have solved some of the atmospheric problems by locating observatories on high mountains, so placing them in thinner regions of the atmosphere. One interesting fringe benefit of this is that astronomers get to work in some of the most beautiful and desolate places on Earth. Even at high altitudes, however, atmospheric interference remains a problem. Most astronomers interested in ultraviolet light, gamma-ray, or x-ray wavelengths do their research from orbiting spacecraft, such as the International Ultraviolet Explorer, ROSAT, the Cosmic Background Explorer and now HST.

When Hermann Oberth outlined his first ideas for a space telescope in the early part of the 20th century, humankind was more preoccupied with preparations for war. The idea of such a telescope was not an important one, and it was readily dismissed to the realm of science fiction. The advent of World War II also turned any thoughts of peaceful astronomical use of the booster rockets (needed to get the telescope into orbit) toward military applications.

At least one other scientist (and science-fiction fan) was interested in the idea of a space telescope. He was astronomer Lyman Spitzer. He spent the war years researching underwater weapons, but kept up an interest in rocket advances and science fiction. In 1946, Spitzer wrote a paper suggesting a space telescope for RAND, a 'think tank' operated by the Douglas Aircraft Company. The paper, which was quickly classified, was called 'Astronomical Advantages of an Extra-Terrestrial Observatory', and it probably served as the 'birth announcement' for the idea of what became the Hubble Space Telescope.

Spitzer's paper was prophetic, because it suggested nearly every field of study that HST is doing – from ultraviolet studies of supernovae and analysis of eclipsing binary stars, to

Figure 6.4. The late Lyman Spitzer, photographed in 1975. *(Lyman Spitzer, Princeton University)*

measurements of the structure of distant galaxies, globular clusters, and quasars. Spitzer did not find his own scientific prescience surprising. As he stated in a conversation a few years before his death in 1997, 'Anybody who was foolish enough to think that rockets might someday be useful for sending things into space would naturally end up with a list very similar to what I wrote.'

In his RAND report, Spitzer recognized that such a telescope would have to be a large and radically different instrument from anything previously used. He wrote:

> Most astronomical problems could be investigated more rapidly and effectively with such a hypothetical instrument than with present equipment. However, there are many problems which could be investigated only with such a large telescope of very high resolving power. It should be emphasized, however, that the chief contribution of such a radically new and more powerful instrument would be, not to supplement our present ideas of the universe we live in, but rather to uncover new phenomena not yet imagined, and perhaps to modify profoundly our basic concepts of space and time.

In May 1990, Appendix V of Spitzer's larger 1946 paper was reprinted in *The Astronomical Quarterly*, and he was invited to contribute his thoughts more than four decades after his original publication. He wrote,

> The chief effect [of this paper] was on me. My studies convinced me that a large space telescope would revolutionize astronomy and might well be launched in my lifetime.

Spitzer spent the 1960s pushing for a large orbital telescope in front of NASA and Congress. His lobbying efforts went on during a period of NASA research and development for the first generation of orbiting telescopes, called the Orbiting Astronomical Observatories (OAOs).

Studies of these orbiting observatories went on against the backdrop of NASA's highly publicized manned space exploration efforts and the headlong rush to put a man on the Moon. It was a heady and interesting time for advocates of a space telescope. In the early part of the 1960s, few things seemed impossible, no matter how technologically complex they might appear. During this time, the telescope – then called the Large Space Telescope (LST) – was by far the most technically complex unmanned project under consideration by the agency. Its crowning glory would be a 3-meter (120-inch) main mirror.

Nancy Roman, who was NASA's Chief of Astronomy at the time, later becoming Program Scientist, with responsibility over a broad range of projects, said of the first official proposal for a space telescope, 'Basically, Spitzer proposed what really was LST back in 1946, but I don't think anybody took it seriously until 1962.'

Roman admitted to being openly skeptical of such a project because she felt that promoters of an orbital telescope were underestimating the engineering complexity of such a mission. There was also the matter of determining just what kind of science could be done with a space telescope. In 1966, Lyman Spitzer chaired the first meeting of a committee from the prestigious National Academy of Sciences looking into the possibility of such a telescope. Called the 'Ad Hoc Committee on the Large Space Telescope', the group studied the range of possible uses for LST, and met with other astronomers to discuss ways to implement the project. In 1969, the committee published a report called 'Scientific Uses of the Large Space Telescope', that gathered all the ideas and plans for LST into several major proposals.

As a member of the Ad Hoc Committee, NASA's Nancy Roman eventually decided that a Large Space Telescope could be built, and she took up the battle for the project. 'I didn't really become a fighter for space telescope until the early 70s,' she said. 'Before that I didn't think that we were ready for a telescope of the size and complexity that everybody wanted, until one of the Orbiting Astronomical Observatories had flown successfully.'

Roman had good reason to be cautious, since she had responsibility for scientific oversight of the Orbiting Astronomical Observatory program. Two successful missions – OAO-II and Copernicus – flew in the 1960s and early 1970s, and made up the first generation of space

Figure 6.5. (Opposite) Edwin Hubble (1889–1953) at the 48-inch Schmidt telescope on Palomar Mountain. *(The Observatories of the Carnegie Institution of Washington)*

telescopes. They were not exactly the intermediate-sized projects called for by Spitzer's committee, but they did provide valuable experience in building orbital telescopes.

By the 1970s, support for a Large Space Telescope was concentrated largely in the astrophysics community, with little invited input from the planetary science community. There was a general feeling that planetary scientists had gobbled up a big share of science money with their probes to the Moon, Mars, Mercury, and Venus. Astronomers who studied stars and galaxies wanted to build *their* big science project. So, although Roman and others saw them as an important constituency for a large telescope, planetary scientists were not too welcome at the planning table at the beginning. 'Certainly the people who were pushing a space telescope were not interested in planets,' explained Roman. 'From where I sat, planets seemed to be a useful component from the beginning. My memory is that I was interested in using it [Space Telescope] for planets long before it became a political necessity.'

Once it became obvious that the project would not fly without a broad base of support, the working groups rallied support from the planetary scientists. However, by that time political support was beginning to fade. The somewhat receptive political climate for projects like LST began to change to one of outright skepticism. The financial and spiritual costs of the Vietnam War affected Congressional willingness to fund 'Big Science' projects, and the staggering costs of the Moon missions did not help matters. Against this backdrop of shrinking budgets and what seemed like the beginnings of national apathy toward big-time space missions, NASA and the working groups of scientists studying the telescope were asking Congress to spend somewhere between $US300 and $US700 million for a 3-meter telescope. To a member of Congress at that time, it must have seemed like the astronomical community was everywhere, asking for money: ground-based facilities, such as the Very Large Array radio telescope in New Mexico, were under construction, and there were constant appeals for funds to upgrade existing observatories. This made the request for money to build LST a rather difficult lump to swallow, and the automatic Congressional response was 'Find a way to make it cheaper'. To do that meant making the telescope smaller, which meant compromising the science that could be done with it. In many scientific circles, the Congressional response, although predictable, was greeted with skepticism and an attitude that if the telescope could not be built as it was proposed, then no good science would come from it.

Still, even with a smaller space telescope, there was the tantalizing possibility that astronomers could get *something* out of the project, so the community looked for ways to keep the project going. To do this, astronomers had to become politically aware and organize grassroots support for the Large Space Telescope. This they did, largely through the leadership and lobbying efforts of scientists like Lyman Spitzer and the Institute for Advanced Studies astronomer John Bahcall.

In 1974, a group of scientists called the Science Working Group for LST agreed to look at the development of a 2.4-meter (94.5-inch) telescope. Downsizing the mirror was painful to consider because it meant a loss of scientific capability, but the result was a space telescope project that engineers and scientists predicted would cost no more than $US300 million through the end of its first year of operation. That was a more realistic amount of money for Congress to consider.

The downsizing required a number of design compromises and scientific 'scale-backs' to keep the program on the funding track, and $US300 million remained the magic number for a long time. After several more years of studies, fine-tuning, Congressional lobbying by members of the science community, and repeated attempts to delete the telescope from the NASA budget, Congress approved an updated budget of $US425–475 million to build the project. By the summer of 1977, the way was finally cleared for the construction of the telescope.

There *was* one catch: in return for the slightly higher price tag, Congress wanted NASA to seek a foreign partner in the project. In 1977, NASA and the European Space Agency signed an agreement setting up a collaboration between the two agencies to run Space Telescope as a joint NASA/ESA program. In essence, ESA agreed to supply 15 per cent of the development cost for the program, making this contribution in the form of the Faint Object Camera and

Figure 6.6. HST on orbit, at the end of the second servicing mission February 1997. *(NASA)*

the solar arrays. In return, ESA astronomers would be guaranteed 15 per cent of available observing time. ESA also agreed to provide some staff members to the yet-to-be-formed Space Telescope Science Institute as its part of the bargain. The stage was set, and it was time to make the telescope a reality. Launch was scheduled for 1983.

(As a short historical aside, it is interesting to trace the line of name changes that the project went through in its developmental years. The space telescope first emerged as 'Large Orbital Telescope', then became the 'Large Space Telescope' (LST). Eventually that was shortened to 'Space Telescope' (ST). In 1983 it was officially named the Edwin P. Hubble Space Telescope, after the US astronomer who first discovered the relationship between the velocity and distance of receding galaxies that describes the expansion of the universe.)

Building HST

How do you go about building a space telescope? The most important thing you need (besides abundant funding) is a mirror. It cannot be just any mirror; it must be a specially made optical system built to withstand the rigors of launch and an orbital deployment in a micro-gravity environment. It needs to be mounted in a spacecraft assembly along with an array of scientific instruments, guidance sensors, communications antennae, solar panels to supply power, an on-board computer, and a variety of other equipment. When that has been accomplished, you test it and then launch it.

This was the process initiated by NASA in early 1977. As soon as the project was approved, the agency faced the gargantuan task of designing the telescope and selecting contractors to build the spacecraft and the instruments that would fly aboard it. The agency also had to determine which NASA center would supervise the contractors, and had to find a home for an institute to run the telescope after it had been built and launched. NASA virtually needed to amass a 'standing army' of people necessary to make HST a reality.

NASA had a complex engineering and scientific undertaking on its hands. Fortunately, there were capable people within the agency to work with contractors and oversee the construction of the telescope. For example, as Program Scientist, Nancy Roman was there to help expedite the process and to make sure the contractors kept the science objectives in mind. 'I felt my job, not only with Space Telescope, but with other projects as well, was to keep the two sides talking to each other,' she explained. 'I used to say when I first joined NASA that I felt I was acting as an interpreter between the engineers and the scientists because, while they both wanted the same thing in the end, they didn't speak the same language.'

Early on in the process of getting the engineers to design and build what the scientists wanted, the NASA Marshall Space Flight Center at Huntsville, Alabama, was selected to be the lead management center for the project. The NASA Goddard Space Flight Center in Greenbelt, Maryland, was chosen to be the operational 'nerve center' for the telescope, and was given responsibility to develop the scientific instruments.

Scientists competed for membership on the Investigation Definition Teams that would define the scientific goals and oversee construction of the instruments. Those instruments were:

- the Wide Field and Planetary Camera (WF/PC-1);
- the Faint Object Camera (FOC) (to be built by the European Space Agency);
- the Faint Object Spectrograph (FOS);
- the High Resolution Spectrograph (now called the Goddard High Resolution Spectrograph or GHRS);
- the High Speed Photometer (HSP).

In addition, plans were laid for servicing missions that would allow scientists to 'swap out' instruments as needed, and for controllers to upgrade any critical systems.

Construction of the Space Telescope began in 1979, with the creation of the mirror and the building of the optical telescope assembly. To form an idea of the extensive network of companies and individuals working on the telescope, consider the following numbers: 21 major subcontractors, one university, and three NASA centers spanning 21 states and 12 other countries were responsible for some aspect of the telescope's components, software design, and other elements – from the mirror down to the batteries, star-trackers, data gathering systems, latches, solar arrays, and a myriad other electronic devices essential to the smooth operation of the telescope. A detailed schematic of HST is shown in Figure 1.13.

Almost from the start, the Space Telescope program ran into financial and political difficulties. Many complications stemmed from managerial entanglements, which in turn arose from the expensive nature of the project and the technical complexity of the telescope. Naturally, when large sums of money are involved in a government project, a cumbersome administration takes charge of spending that money and directing the actual work. It takes time for commands to travel from the top of the hierarchy to the bottom in such a bureaucracy, and, furthermore, HST was built under tremendous deadline pressure.

Time pressures do not exist gracefully alongside technological development. Nearly every group involved in designing and building the telescope ran into difficulties in design and construction. These stretched out the time it took to build each of the telescope's intricate mechanisms. Space Telescope was a first-generation design, and new technologies had to be developed for instruments such as the Fine Guidance Sensors. The design of the entire optical telescope assembly was part of what NASA managers called 'advanced development'.

Designing and building new technology costs more than using tested designs, and many items – computer memories, for example – change rapidly. Consequently, what worked for previous space vehicles became out of date and inappropriate for HST, leading to the necessary development of new technology. So it went with nearly every part of the telescope, and the budget continued to expand. By 1985, the cost of the telescope was estimated to be $US1.175 billion, which was just enough to cover assembly and the on-orbit verification program. By 1986, that figure had grown to $US1.6 billion just to design and develop the telescope. That was the year the *Challenger* exploded, postponing HST's launch until 1990.

The shuttle tragedy may have led to unexpected benefits for the telescope, since it still was not ready for launch in 1986. Engineers took advantage of the delays to make improvements to systems on board the spacecraft, and to test others to make sure they were still in working order. For the Faint Object Spectrograph, the lengthy launch postponement meant that part of a critical mirror oxidized while in storage. This reduced the reflectivity, a problem which was not discovered until after the telescope was launched.

Storage expenses and costs of testing and re-testing all had to be added into the budget. By the time the telescope was finally launched into space, had operated for three and a half years, and was refurbished, it cost $US2.3 billion. At the time of writing, the total yearly budget for HST is in the range of $US250 million a year. Looking at it another way, the total telescope costs nearly $US8.00 per second to use!

Politically, the Space Telescope program was a maze of turf battles. Contributing to the general difficulty of building the telescope were the adversarial relationships between NASA agencies and contractors. Marshall Space Flight Center, which had overall responsibility for the contracts to design and build the telescope, fought with Goddard Space Flight Center, which wanted control of the project. NASA Headquarters' viewpoints conflicted with the European Space Agency's goals. The Instrument Definition Teams – groups of scientists charged with the responsibility of designing the on-board instruments – fought with the newborn Space Telescope Science Institute. The Space Telescope community itself received criticism from ground-based astronomers and others who felt that precious science money was going into one of the largest boondoggles ever funded by NASA. Even other space observatory teams felt left out.

To defend HST from continual assaults and budget criticism, NASA put pressure on contractors to cut costs and boost efficiency. Designs changed and costs were cut, but the telescope suffered each time this happened. In the case of Perkin-Elmer and the mirror grinding, pressure made a bad situation worse. The company, which had underbid the project at $US69.5 million so it would be chosen to do the job, was running way behind schedule in its attempt to provide a finished mirror. Costs were mounting, and the company managers told NASA that the project would cost more than they had bid. NASA agreed to pay more after a review of the process, but urged Perkin-Elmer to finish the job quickly. The pressure was on the company to perform – so, in the interests of saving time and money, crucial tests of the mirror were omitted. The tests performed by Perkin-Elmer were inadequate, and, as we saw earlier, incorrect. The company finished polishing the mirror in 1981, claiming it exceeded NASA's specifications. Three years later, it delivered the completed Optical Telescope Assembly to NASA and presented a final bill for more than $US300 million, around five times the original bid.

The Space Telescope Science Institute

From the start, it was clear that the astronomy community did not want NASA to coordinate the use of the telescope. Historically, the relationship between scientists and NASA has been a wary one, and nowhere was it more contentious than during planning for the Space Telescope Science Institute. In the early years of the space agency, the National Academy of Sciences had formed a Space Science Board to recommend future policy. NASA managers were not happy taking orders from an outside group, and wanted to use their own scientists to make decisions. This did not sit well with those outside the agency, and, in any case, it just was not good politics for NASA to ignore the viewpoints of those working elsewhere.

The concept of forming an institute to implement the science program of the Space Telescope first came to light in a 1972 report issued by HST Project Scientist Bob O'Dell. It was assumed that Goddard Space Flight Center would be the science operations center as well as the control center for the telescope, but that assumption was challenged by various members of the science community. At stake were large amounts of telescope time on what was anticipated to be the premier observatory in astronomy, as well as large sums of money to be spent on the observatory operations. To give just a couple of examples of the scientists' concerns: some felt that NASA scientists would 'skim the cream' of observing time on a space telescope, depriving outsiders of access; there was also a feeling that was stated generally as 'NASA is so big, it'll just mess this up'. Obviously, some measure of independent input and oversight was needed, so another working group was formed to look into ways in which an institute could be formed. That committee looked at every possibility, ranging from giving NASA complete responsibility, to finding an outside consortium to run the telescope.

The stage was set for the development of the Space Telescope Science Institute. After a competition among several consortia, the Associated Universities for Research in Astronomy (AURA) group was selected in 1981 to run the Institute at Johns Hopkins University in Baltimore, Maryland. Computer Sciences Corporation (CSC) would support the controllers, managers, programmers, and other professionals needed to administer the telescope operations.

Now all that was left was to launch the telescope and start observing with it.

Launch, deployment, and 'attitude shift'

No one who has ever seen a space shuttle launch can adequately describe the intense sights and sounds they experience. Large-screen IMAX movies come the closest to recording the event, but there comes a point when no movie can duplicate the Earth-shaking power of a shuttle leaving the ground and moving to orbit in less than 10 minutes. People attending a launch watch it in very different ways. Some cluster in groups at VIP sites around the center. Others watch from the press stands, and thousands gather along causeways and highways leading into the Kennedy Space Center to experience a shuttle blasting off. Some choose to watch the launch alone, away from the crowds. Others hold hands and hug as the shuttle thunders its way into the sky.

In late April 1990, HST was launched as thousands watched. NASA's announcer saluted the mission: '...and liftoff of the space shuttle *Discovery* with the Hubble Space Telescope – our window on the universe.' A day later, the telescope was deployed in orbit over South America, and people all over the world watched as HST's solar arrays were unrolled and the telescope was set free. For the scientists and the teams watching the event, it was the realization of a dream. It was an emotional moment. Some laughed; others cried. All were acutely aware that it was the beginning of a scientific adventure they would never forget. It just was not the adventure they originally planned on having.

In a roundabout way, this brings us back to the mirror problems, the jitter, and the legion of other difficulties that people had with HST. Where in that history of technical oversight, painstaking planning, and innovative design did HST go wrong? There are some who say that the mirror was ground incorrectly because there was not enough money to do it right. Others point out that the opticians at Perkin-Elmer were under a lot of pressure to do the job right, but to do it fast. Recall that the Perkin-Elmer people improvised during the measurements of the mirror's surface 'fingerprint'. If they had not been under the gun to produce a working mirror for NASA, perhaps today's HST would not have spent three years as 'the hobbled Space Telescope'.

Just as HST's reputation has changed since launch, so too have the human members of HST's team. For most of the scientists who lived through the dark days of summer 1990, it was a roller coaster of emotions. They went from the highs of launch to the lows of tragedy and grief when the spherical aberration was discovered. Looking back over that time, WF/PC team member Sandra Faber wrote,

> It was the most frantic, most exhilarating, yet most frustrating time of my life. We were working 100 hour weeks, and when I fell into bed at one a.m., I could not sleep. I was turning over a million theories in my mind, plus a million strategies for getting new data and analyzing it. I was on a continuous adrenalin high. This lasted the whole summer, well into fall.

If not for the efforts of hard-working and creative people, the HST might well have continued to be thought of as a complete failure. When some within NASA and the science community sought to write it off, team members pulled together and came up with ways to salvage the impaired telescope. What changed? Who can say? Perhaps it was that all-too-human propensity to unite in the face of adversity.

Faber, who still works with HST data, remembers watching attitudes shift between factions within the telescope community. 'The Space Telescope was a troubled project from the start,' she noted. 'It was full of little groups that had nothing to do for years except fight with one another while waiting for the telescope to go up.'

Somewhere between the tense turf battles during the telescope's formative years and the ugly finger-pointing that went on after Hubble's mirror problem was discovered, Faber believes there was '... an about-face'. She continues, 'People really began to pull together. They forgot their animosities and got on with the job of understanding a very complex and difficult project that was in deep trouble. There's the nice human side of the story in all this.'

Scientists' perceptions about HST have changed, too. Despite the difficulty of receiving data from the telescope during the first three years, and the hassles inherent in 'deconvolving' the data, astronomers have lined up six deep to use this twice-refurbished and updated observatory. HST today, in fact, is hugely oversubscribed – which is a testament to the good science it can do.

Part of HST's human story is the rapid and united effort that scientists and engineers made to design repairs for HST's faulty mirror. In particular, the Space Telescope Science Institute played a vital role in pushing for a refurbishment mission. The day after spherical aberration was announced to the world, Space Telescope Science Institute astronomer Holland Ford began pressing for ways to repair the optics. After some months, an HST strategy panel was formed. Ford and another Institute astronomer, Bob Brown, chaired the panel, which included Lyman Spitzer and Jim Crocker. The panel's charge: to find a 'fix' for HST's mirror.

According to Spitzer, all possible ways to deal with the mirror were up for consideration. There were even proposals to bring the telescope back to Earth, a dangerous technical undertaking: 'We knew that some highly placed people were pushing for bringing the telescope back down,' he said. 'One of the main arguments that we had against it was a political one – that once we got it down, could you ever get it back up again?'

The Corrective Optics Space Telescope Axial Replacement (COSTAR) was the most palatable idea proposed to the panel. Basically, COSTAR was a device that would take the light coming from the aberrated mirror and 're-focus' it for use by the spectrographs and the Faint Object Camera. However, the incorporation of COSTAR required removing the High Speed Photometer, which meant expanding an already crowded servicing mission schedule. Nevertheless, the panel presented the COSTAR proposal to NASA, which evaluated the idea and approved it.

The 'fix' for the Wide Field and Planetary Camera was deceptively simple. It involved a 'clone', called WF/PC-2, which was intended as an emergency replacement should the original fail. Building the clone was underway, although it was not a high-priority development at the time. After a meeting of the repair strategy panel, members of the 'Clone Team' pointed out that the WF/PC-2 could be retrofitted to account for the effects of spherical aberration. To do this, one mirror in each optical quadrant of the replacement WF/PC could be refigured. At that point, interest in the WF/PC development mushroomed, and the project was accepted for the first servicing mission.

Memories of missions

Everyone has memories of 'defining moments' in the history of the Hubble Space Telescope. For some scientists, those moments will take place during hallway conversations and paper sessions at professional meetings for years to come. Others will experience their moments when they glimpse a distant quasar or galaxy, frozen in an HST image for an instant in time. For others, it will be the first time they take spectra of a galactic jet and realize that they have indisputable evidence of another black hole.

For many members of the public, those defining moments came while watching the First Hubble Servicing Mission unfold on television. HST team members converged on the Kennedy Space Center in Florida to watch one of the most anticipated missions in the history of manned space exploration. What they got was one of the best-planned and beautifully executed shuttle missions ever flown.

The sounds and images of astronauts working in space made the First Servicing Mission a very exciting and dramatic television experience. For 11 days, Story Musgrave, Jeff Hoffman, Kathryn (K.T.) Thornton, and Tom Akers entranced the world with their extra-vehicular repair expertise, while Dick Covey, Ken Bowersox, and Claude Nicollier co-ordinated the mission from inside the shuttle. They had 11 tasks to accomplish, and they were so well-trained that they made a tough job look easy. In the process of repairing HST, they also went a long way toward repairing NASA's tarnished public relations image.

After the exhilaration of launch and the servicing mission came the hardest wait of all – waiting to see if HST had truly been repaired. HST Project Scientist Dave Leckrone, in a press conference shortly after the mission, stated that the repair teams had 'finished eye surgery on the patient'. He expected that there would be a six to eight-week wait until controllers could take off the bandages and test the patient's eyesight. Everyone waited while controllers learned to use the newly refurbished instrument and began the ticklish process of deploying the corrective optics.

For a small group of scientists who were lucky enough to be at the Space Telescope Science Institute very early one mid-December morning in 1993, their memories are forever linked to the moment when the 'first-light' image from the newly installed WF/PC-2 appeared on a computer screen. To the uninitiated eye, it looked like a hash of pixels, but to the jubilant scientists crowded around the monitor it was a *focused* hash of pixels. After three and half years of frustration, hard work, training, and planning, HST was capable of doing all the science it was sent to do, and more.

In 1997, astronauts once again visited HST to make repairs and upgrades. Over the course of a smoothly run, 10-day mission, the crew removed the Goddard High Resolution Spectrograph and the Faint Object Spectrograph and replaced them with the two 'second-generation instruments' – the Space Telescope Imaging Spectrograph (STIS) and the Near-Infrared Camera and Multi-Object Spectrometer (NICMOS) both of which were described in Chapter 1. Since their installation, these two instruments have been surveying such objects as the expanding ring of gases around Supernova 1987a (discussed in Chapter 3), and probing the interiors of supermassive black holes. Unfortunately, the NICMOS instrument developed a severe problem right after the servicing mission – one which has affected the observation schedule of the entire telescope. In order to do its infrared-sensitive surveys, NICMOS uses cryogens to keep its detectors cooled. The instrument was installed with approximately 4 years' worth of coolant. However, a thermal short developed in the assembly – called a dewar – that holds the coolant. This allowed a loss of coolant, which shortened NICMOS's useful lifetime to about 19 months. In order to do as much NICMOS science as possible before the coolant runs out, a variety of other observations have been put on hold as NICMOS forges ahead.

There may be a way to repair the NICMOS coolant problem. Scientists and engineers are studying a high-tech heat pump cooling system for the instrument – the so-called NICMOS Cooling System. If feasible, the unit will be installed during the third HST servicing mission.

Figure 6.7. Scenes from HST's second servicing mission, February 1997: **(previous page)** a spectacular night launch; **(above)** HST just before it was captured for servicing; **(right and overleaf)** astronauts working on the telescope; **(page 206)** landing at Kennedy Space Center at the conclusion of the successful mission. *(NASA)*

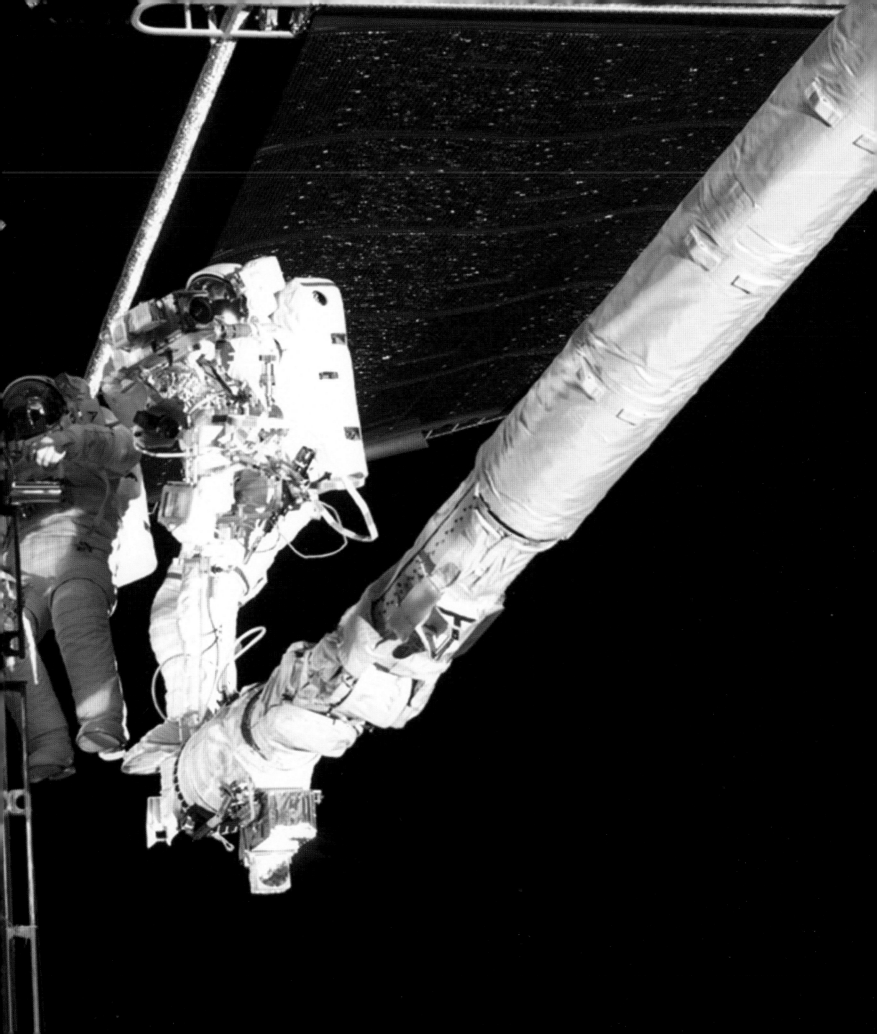

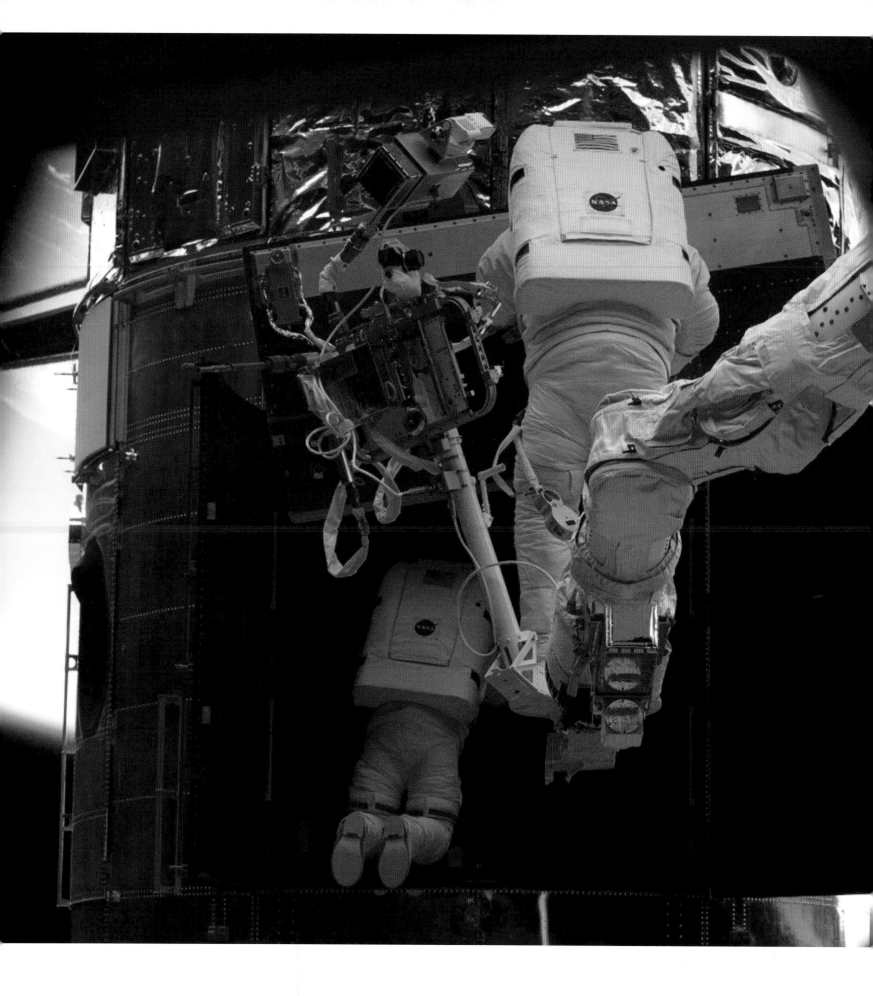

The once and future Space Telescope

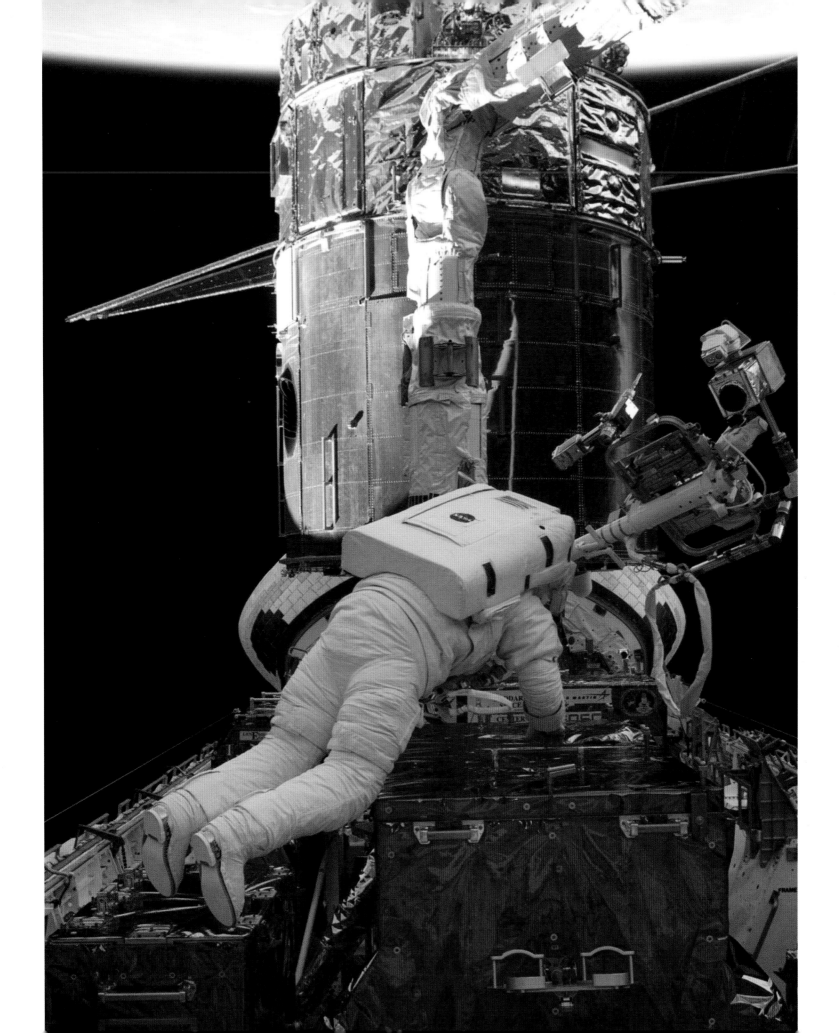

The Hubble end game

The Hubble Space Telescope's work continues, with more than 400 observation programs scheduled each year. The next servicing mission is set for late 1999. During that time, astronaut crews will install the Advanced Camera for Surveys, and may attempt to make the needed repairs to the NICMOS instrument. If a repair is possible with the 'cryo-cooler' project, it will be accomplished before the telescope is re-boosted to a higher orbit. After that time, a number of highly significant observations will be scheduled for completion, along with the normal run of observation programs.

The Advanced Camera for Surveys (ACS) will be installed in the axial bay of the telescope where the Faint Object Camera was installed. ACS will widen the imaging capabilities of the telescope. Users will be able to choose between a 200 by 200 arcsecond wide channel optimized for deep imaging in the optical and near-infrared, a high-resolution mode offering near-UV and optical imaging, and a high-throughput far-ultraviolet channel. The principal investigator for ACS is Holland Ford of Johns Hopkins University.

HST's final scheduled servicing mission is set for 2002, and new instruments will be installed. The first is called the Cosmic Origins Spectrograph (COS), being developed by a University of Colorado team headed by James Green. COS will enable its users to dig deeper into such cosmic mysteries as distant quasars, and further map the Lyman-alpha forest (discussed in Chapter 5). Of particular interest to astronomers are starburst regions in galaxies, and COS should provide an excellent view of these high-energy areas of the universe. The second instrument will probably be a low-cost new Wide Field Camera (WFC-3). Currently, no servicing missions are planned after 2002.

The original end of HST's mission was set for 2005, but there are good signs that the telescope that has served astronomers so well will be around for a bit longer. Robert Williams describes the last days of Hubble as a time of discovery and reassessment. 'We are doing a review of HST's "end game",' he said, 'and the question I like to ask is, "how do we maximize discovery space with HST?"'

How indeed? It is clear that Hubble is a unique facility. It fuels not only discovery in outer space, but a tremendously large research effort on Earth. Each year millions of dollars are spent on the analysis of Hubble data; this has had a tremendous effect on the infrastructure of astronomy research in the United States. With reduced funding and a wind-down of activity by the year 2010, some things will have to change. With the installation of new instruments during the final servicing missions, it is clear that the telescope can remain useful well past its nominal 2005 end date, and even beyond the end of an extended mission in 2010.

Williams pointed out that the best question to ask now is, What is the best use of Hubble with the final instruments it will receive? 'Maybe it would be best to operate Hubble in more of a survey mode and give more time to large projects that systematically try to attack problems and gather uniform data,' he said. 'Or maybe we should emphasize ultraviolet astronomy and spectroscopy for example, because it's an important part of astrophysics.'

That suggestion is an important one because, even though NASA has Far Ultraviolet Spectroscopic Explorer (FUSE) on the horizon, it does not have the same capabilities as HST for wide-field high-resolution spectroscopy. Aside from the prospect of the COS instrument, there is no major ultra-violet spectroscopic facility planned for the foreseeable future. It is possible

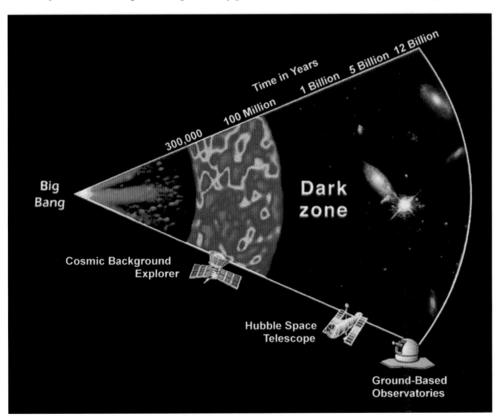

Figure 6.8. Our current understanding of the universe begins with the Big Bang and subsequent evolution of the cosmos to the countless numbers of galaxies and stars we see today. The Next Generation Space Telescope (NGST) will look back to the time when the stars and galaxies were just starting to form – the so-called 'Dark Zone' – a time beyond the reach of current instruments. *(P. Stockman, STScI; NASA)*

(a)

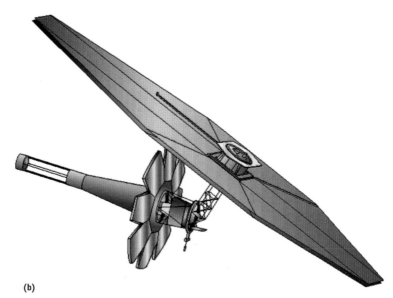

(b)

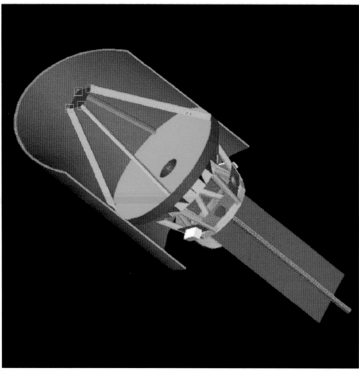

(c)

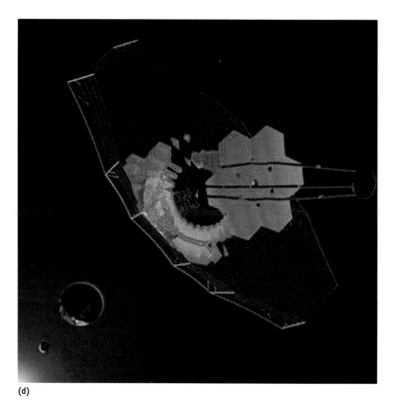

(d)

Figure 6.9. The space telescope of the future? Designers are already at work on concepts for the Next Generation Space Telescope (NGST). It would orbit at a Lagrange point, which would allow for passive cooling, and easier orbital operation. The instruments on this telescope would emphasize infrared studies. *(Concepts courtesy (a) Ball Aerospace Corporation; (b) NASA-Goddard Space Flight Center; (c) TRW Corporation; and (d) Lockheed Martin Corporation)*

that, after Hubble, ultra-violet astronomy could lay fallow. Williams suggested that, in anticipation of a reduced-ultra-violet capability, a re-boosted future HST could emphasize ultra-violet astronomy throughout the last years of its useful lifetime. It could even complement the Next Generation Space Telescope, if the two observatories should overlap their lifetimes. The current plan is to de-orbit HST in 2010.

Intimately tied to HST's end game is the future of the Space Telescope Science Institute. Once HST is no longer on orbit, or functioning as a working observatory, the wealth of information and experience that is represented by the Institute could well be dissipated into the community. To prevent that loss, Williams and others are reassessing the function and future of the Institute in a post-HST world. Williams, who passed the directorship of the Institute on to Steven V. Beckwith on September 1, 1998, stated that the capabilities of the facility could be used to support future missions, or perhaps to broaden its outlook and enable it to become a center for research in astronomy and astrophysics. Beckwith, a seasoned astronomer with a background in infrared observing, is well-positioned to maintain STScI's role in research with future space telescopes. The announcement in June 1998 that the Institute would assume management of science operations for the Next Generation Space Telescope was an expected 'next step'. Certainly in its history STScI will continue to maintain and analyze HST data, and extend its active education and public outreach programs in astronomy.

The odyssey of Hubble Space Telescope has been a long, exciting, and sometimes depressing one. It has gone from being a scientist's dream to an expensive 'Big Science' project in a little over 40 years. During that time, HST has been many things to many people – a multi-billion-dollar boondoggle, a complex engineering project, a failure, a 'techno-turkey', a miracle machine, a job, an obsession, a whipping boy, the future of NASA, the death of NASA, the solver of mysteries, and the key to the universe. HST's future is much brighter now than when it was first launched, but it is important to remember that the telescope *did* achieve impressive science even during its difficult first years. Proof of HST's success comes from increased public acceptance of the telescope and its accomplishments.

The scientific rewards that HST promised are at hand. In 1996, HST took its 100 000th exposure – a milestone that some thought would never come. The newly improved telescope is often more productive than the most productive ground-based telescopes with which it works. With proper care and maintenance, it could last well into the first decades of the 21st century.

It is an exciting and rewarding time to be an astronomer. The less tangible rewards are found in the resiliency of the human spirit – in the determination that scientists, engineers, politicians, and administrators showed in the face of continuing adversity throughout the program. Ed Weiler, in an interview before the First Servicing Mission, said: 'My greatest reward on this program will be able to get in front of my friends in the media who turned out not to be so friendly and remind them that three and a half years ago when we said we had a problem, we also told them we could fix it by the end of 1993. Three years ago, HST was called a techno-turkey, a disgrace, a dead satellite. But, we worked very hard to get science out early and fast. It is a tribute to everyone's hard work that Hubble has generated more positive press stories in the last two years than all of the NASA satellites combined.'

Right now, the most tantalizing cosmological work to come from HST's data is the study of the earliest epochs of the universe. With HST and ground-based instruments, we can see back to a time when the universe was a small fraction of its current age – perhaps when it was only 3 billion years old.

The questions raised by HST's pioneering studies will almost certainly be answered by the Next Generation Space Telescope (NGST). It will serve a new generation of astronomers who specialize in objects that radiate much of their light in the visible and far-infrared portions of the electromagnetic spectrum. NGST's primary mission will be to look beyond the 3 billion year old limit and image the first galaxies and supernovae in the universe, and perhaps answer the question posed by the discovery of fragments of galaxies in the Hubble Deep Field: How did the current structure and order of the universe come into being? Closer to home, this new telescope will search for planets around other stars – a stimulating quest for anyone who looks out at the night sky and simply wonders about the possibilities for life elsewhere in the universe.

The history of astronomy is an ever-repeating cycle of expanding horizons and technological advancement. From Galileo's first telescope, which improved on the naked-eye vision of the universe, to the complex assemblages of optics and electronics we use today, astronomical discovery has risen – and fallen – on the strengths of the equipment used to achieve it.

HST's age of discovery can be likened to Galileo Galilei's astronomical accomplishments. Today, there are hundreds of Galileos, pointing the Space Telescope in all directions, finding mind-boggling things. The discoveries will continue for the foreseeable future, as long as the telescope remains healthy, and – more importantly – as long as there are questions to be answered about those things which most fascinate us about the universe.

Glossary

In addition to the usual, short entries, this section also includes extended discussions of physical concepts helpful in understanding some of the research reported. Italics denotes other entries in the glossary.

absolute temperature scale In the physical sciences, temperatures are related to the motions of atoms and molecules in gases. The temperature scale used is the Kelvin scale, which has the same size degrees as the centigrade scale. On the Kelvin scale, the freezing point of water is 273 kelvin, the boiling point of water is 373 kelvin, and the surface of the Sun (*effective temperature*) is 5750 kelvin. The temperatures in this book are all given in kelvin.

absolute zero A hypothetical concept to describe the temperature, 0 kelvin, at which the motions of atoms and molecules would stop.

absorption lines When a continuous spectrum is viewed through a cooler, low-pressure gas, such as the interstellar gas, dark lines called absorption lines appear. The specific lines depend on the composition of the gas that is absorbing the light and causing the lines.

active galaxy A galaxy with a bright central region and often with enhanced x-ray, ultraviolet, and radio emission. Active galaxies include the Seyfert galaxies and *quasars*. The activity probably results from a massive, central *black hole*.

angles Angular measure is frequently used in astronomy to specify positions on the sky and to specify apparent size. The units are degrees, minutes, and seconds. An entire circle contains 360 degrees; a right angle contains 90 degrees. Each degree contains 60 minutes, and, in turn, each minute contains 60 seconds. Note that the minutes and seconds described here are not the same as minutes and seconds of time.

angstrom A unit of length equal to 10^{-10} meter.

arcsecond A commonly used, very small, angular measure in astronomy. Roughly speaking, it is one part in 200000. To illustrate the size, a US quarter coin is about 2.5 centimeters across. If placed at a distance of 200000 quarters (about 500000 centimeters i.e. 5 kilometers), it subtends 1 arcsecond. If we are talking about 0.1 arcsecond, it is about the angular size of a quarter 50 kilometers distant.

asteroids Sometimes called 'minor planets', these are small, rocky bodies orbiting the Sun primarily between the orbits of Mars and Jupiter, the region of the 'asteroid belt'. Diameters range from the smallest detectable (at a few tenths of a kilometer) up to the largest asteroid, Ceres, at 950 kilometers.

astrometry The measurement of precise positions and motions, usually of stars.

astronomical unit (AU) The average distance between the Earth and the Sun, i.e. about 150 million kilometers.

atomic number The number of protons in the nucleus of an atom (and hence the number of electrons in a normal, unionized, atom).

atomic weight Approximately the number of protons and neutrons. For example, a helium atom, with two protons and two neutrons, has an atomic weight of 4.003.

baryons A class of elementary particles that includes the *proton* and the *neutron*.

Big Bang The postulated beginning of our universe when all the matter and radiation emerged from a point.

Big Crunch If the universe should stop expanding and fall back on itself, it eventually could concentrate all the matter and radiation back into a point. This event is called the Big Crunch.

binary stars Two stars in orbit around each other. If the separation can be resolved in the telescope, they are called visual binaries. If the double nature is revealed by motions detectable via the *Doppler effect*, they are called spectroscopic binaries. Most stars are believed to be members of binary or multiple systems.

black body An ideal body which absorbs all radiation incident on it. The emission from a black body depends only on its temperature.

black hole When massive stars collapse at the end of their lives, their mass is concentrated at a single point. The effective size of the star is called the *event horizon*, the boundary of a region with a gravitational field so strong that nothing can escape from it, not even light. Hence, the name 'black hole'. The lower limit for the mass of a black hole is about 3 solar masses. Of course, this is the mass of the final core, not the original mass of the star on the *main sequence*, which can be much larger. Black holes with other origins are possible. So-called 'mini black holes' may have formed early in the universe, and 'massive' black holes, with masses of millions to billions of solar masses, are probably located at the centers of galaxies.

blast wave A traveling *shock* produced by an explosion.

carbon cycle Nuclear reactions that convert four hydrogen nuclei into one helium nucleus with a release of energy using carbon nuclei as a catalyst. The carbon cycle occurs at higher temperatures than the *proton–proton chain*.

catalog numbers Many astronomical objects in this book are designated by their catalog number. For example, M31 means the 31st object in Messier's catalog, and NGC 188 simply means the 188th object in the New General Catalog. Others found in this book are Arp numbers, CL numbers, Henize numbers, Markarian numbers, Melnick numbers, PKS numbers, and 3C numbers. These are from specific catalogs or lists, often named after the astronomer compiling them. While we give the numbers for completeness, they are not essential for understanding their role in HST science.

Cepheid variables Very luminous, pulsating stars that are important distance indicators. Their brightness pulsates regularly over a length of time called a *period*, and their *intrinsic brightness* can be determined from the period of light variation.

cluster parallax A distance determination method applicable to clusters with measurable motions toward a convergent point (where all the motions would intersect on the sky) and radial velocities determined by the *Doppler effect*. The distance to the Hyades *Open Cluster* has been determined using this method.

clusters of galaxies Distinct groups of galaxies containing from ten to thousands of galaxies.

clusters of stars See *globular clusters* and *open clusters* .

comets Solar System bodies consisting of an icy nucleus. If the comet approaches the Sun, the nucleus is heated, and the cometary ices sublimate to form a cloud of gas and dust called a coma. Sometimes the gas and dust will stream out away from the comet, pushed by the solar wind (for the gas) or the Sun's *radiation pressure* (for the dust) into a long tail.

constellation One of the arbitrary groupings of stars in the sky, of which there are more than 80, which people imagine to look like objects, such as Orion the hunter in the winter sky (northern hemisphere). The brightest stars in each constellation are designated in order of brightness by a Greek letter in alphabetical order; for example, Alpha Orionis, Beta Orionis, etc.

continuous spectrum An incandescent gas under high ppressure (such as the sub-surface layers of the Sun) emits a continuous spectrum, i.e. emission at all wavelengths over a wide range of the electromagnetic spectrum.

cosmic abundances A standard tabulation of the relative abundances of the elements in the universe compiled from solar, meteorite, stellar, nebular, and Earth-crust data.

cosmic background radiation The nearly isotropic surviving radiation from the *Primordial Fireball*. It has a *black body* temperature of 2.73 kelvin.

cosmology The study of the entire universe considered on a very large scale; this includes the origin, structure, and evolution of the universe.

critical density and Ω_0 The density of the universe that would just stop the expansion of the universe after a very long time is called ρ_0. For a *Hubble Constant* of 80 kilometers per second per megaparsec, this amounts to 1.2×10^{-29} grams per cubic centimeter, a very small number. If this were all hydrogen atoms, their density would be 0.7×10^{-5} atoms per cubic centimeter. A more convenient quantity is Ω_0, which is the ratio of the actual density to the critical density. For an actual density equal to the critical density, Ω_0 is 1. For Ω_0 higher than 1, the expansion stops and becomes a contraction, ultimately leading to the *Big Crunch*. For Ω_0 lower than 1, the universe expands forever.

dark matter Unobserved, and hence 'dark', matter in the universe. This matter could be of a straightforward nature, such as very subluminous stars, and/or it could be exotic subatomic particles. The evidence for large amounts of dark matter in the universe is extensive.

deceleration parameter A measure of the rate at which the expansion of the universe may be slowing down because of the gravitational attraction of its own mass.

deconvolution The sharpening of a degraded image (or spectrum), usually by computer processing. If the degradation can be accurately determined, for example by determining the degraded appearance of a star which should be nearly a point source, it can be substantially removed.

deuterium A heavy hydrogen atom containing one proton and one neutron in its nucleus.

Doppler effect A change in the wavelength of light caused by a relative motion of a source and its observer. This change in wavelength can be understood by thinking of light as a wave. If the source approaches the observer, more waves per second are seen; the frequency is therefore raised, and the wavelength is shortened: this is referred to as a blueshift – a shift toward the 'blue' end of the spectrum. When the source is moving away from the observer, the opposite occurs; this is referred to as redshift. We hear the same effect in the sound of a siren; the pitch (frequency) is higher (corresponding to a lower wavelength) when the vehicle approaches us, and the pitch is lower (corresponding to longer wavelength) when the vehicle is going away from us, so, as it passes, a distinct change in the pitch of the siren is noticed.

effective temperature The temperature that an ideal body, called a *black body*, would have if it were to emit the same amount of energy per unit area. The term is usually applied to stars; for example, the Sun's effective temperature is about 5750 kelvin.

electromagnetic spectrum (**EMS**) The entire spectrum of all forms of electromagnetic radiation, including light, from gamma rays to radio waves.

electron The light, negatively charged particle in the atom.

emission lines Incandescent gases at low pressure, such as a gaseous nebula, produce a spectrum composed of individual bright lines called emission lines. The concept of 'line' means that the emission occurs at a specific wavelength or narrow range of wavelengths, a situation in contrast to the *continuous spectrum*. The specific lines depend on the composition of the gas. See also *absorption line*.

event horizon The surface, or effective boundary, of a *black hole*. No material or light can escape from within the event horizon. The radius is approximately 3 kilometers times the mass of the black hole (given in solar units).

fission The break up of atomic nuclei into lighter nuclei.

fusion The merging of atomic nuclei into heavier nuclei.

galaxy One of the very large celestial objects consisting of stars (up to roughly a trillion, 10^{12}) and often vast quantities of dust and gas. The Sun is in the *Milky Way Galaxy*. External galaxies are often described by their appearance, such as spirals, barred spirals, ellipticals, and irregulars.

general relativity Albert Einstein's generalization of *Newtonian mechanics*, which is used in cosmological applications.

giant A star roughly 100 times the Sun's intrinsic brightness with a radius of roughly 100 times the Sun's radius.

globular clusters Tightly packed, spherical groupings of old stars in the *Milky Way Galaxy*. Globular clusters are also observed in other galaxies.

gravitational lensing The general relativistic effect whereby the enormous mass of celestial objects changes the path of light passing by. The effect is to focus the light and thus the object causing the effect is, in essence, a gravitational lens.

gravity The attraction of all bodies in the universe for all other bodies. Two bodies attract each other with a force proportional to the product of their masses and inversely proportional to the square of the distance between them. For example, the solar gravitational force acting on the Earth would be one-quarter the present value if the Earth were twice as far from the Sun as it is now.

H–R or **Hertzsprung–Russell diagram** A display of stellar properties using a plot of *effective temperature* (or surrogates such as color or spectral type) versus *luminosity* (or surrogates such as *intrinsic brightness*).

Hubble Constant, H_0. The constant of proportionality between the recession speeds of galaxies and their distances from each other. Current estimated values range between 50 and 100 kilometers per second per *megaparsec*.

inflation A hypothetical period of extraordinarily rapid expansion early in the universe, from roughly 10^{-34} to 10^{-30} seconds after the *Big Bang*. This expansion may be necessary to explain properties of the universe, such as the existence of galaxies.

interstellar medium The gas and dust located between the stars in the Milky Way Galaxy.

intrinsic brightness The brightness of an object, such as a star, that is independent of distance. This brightness can either refer to light in a specific color or to all the light, when it is the same as the *luminosity*.

ion An atom or molecule that has lost one or more *electrons* and thus has an electrical charge.

isotope Atoms of the same chemical element, but with different numbers of neutrons in the nucleus.

kiloparsec 1000 *parsecs*, or 3.1×10^{16} kilometers.

Kirchhoff's laws Experimentally determined ideas based on the absorption and emission of light. See *absorption line, continuous spectrum,* and *emission line*.

light-travel time The time it takes for light, traveling at about 300 000 kilometers per second, to travel a certain distance.

light year The distance that light travels in one year at about 300 000 kilometers per second, i.e. 9.5×10^{12} kilometers.

look-back time Because light travels through space at a constant speed (300 000 kilometers per second), it takes a finite time to travel from distant objects. Hence, we 'see' distant objects at a point in time in their past; this point in time is the look-back time. See *light-travel time*.

luminosity The total light output of a star over all wavelengths. Because this quantity is independent of distance, it is an *intrinsic brightness*.

Lyman-alpha The principal light-absorbing or -emitting energy transition of the hydrogen atom occurring at about 1216 *angstroms*. Often abbreviated as Lyman-α.

Magellanic Clouds Two irregular galaxies, the Large Magellanic Cloud (LMC) and the Small Magellanic Cloud (SMC), easily visible to the unaided eye in the southern hemisphere. They are named after explorer Ferdinand Magellan.

magnitudes Stellar brightness is often described in terms of a historical system that seems confusing. The magnitude system is somewhat like standings in sports leagues where the best (and brightest) are first, the lesser (or fainter) are second, third, etc. Historically, the brightest stars (collectively) in the sky are first magnitude, and the faintest stars visible to the naked eye are sixth magnitude. The modern definition is that stars differing by a factor of 100 in brightness differ by 5 magnitudes. This works out to a factor of 2.512 in brightness for 1 magnitude, and the fainter stars have the larger magnitude. The magnitudes as observed for stars on the sky are called apparent magnitudes. If the magnitude is referred to the standard distance of *10 parsecs* (and thus is indicative of the star's *intrinsic brightness*), it is called an absolute magnitude. Finally, the magnitude can refer to all the light from a star or to the brightness of light measured through a specific filter, such as a blue filter.

main sequence The principal band of stars on the *H–R diagram*. Most stars appear on the main sequence after nuclear burning of hydrogen has begun, and they spend most of their lives there.

megaparsec One million *parsecs*, or 3.1×10^{19} kilometers.

Milky Way Galaxy Our own galaxy, consisting of about 100 billion stars plus gas and dust. The galaxy is disk-shaped; when we see the Milky Way in the sky, we are looking at the disk edge-on.

nebula A cloud of gas and dust in a galaxy, such as the Orion Nebula. Before the early part of the 20th century, this term was used to describe any hazy patch in the sky.

neutrino A stable particle with no charge and no mass that is produced in nuclear reactions. The neutrino interacts very weakly with all other particles.

neutron The heavy, electrically neutral elementary particle in the nucleus of an atom.

neutron star A very dense star composed of *neutrons* and having a diameter of about 30 kilometers. The pressure in the star is so great that *electrons*, which usually orbit the nucleus of atoms, are pressed into the nucleus, where the protons and electrons merge to form neutrons. The mass of neutron stars falls between 1.4 and 3.0 solar masses. Smaller stars become *white dwarfs*, while larger stars become *black holes*. Neutron stars are believed to result from *supernovae*. *Pulsars* are believed to be rapidly rotating neutron stars.

Newtonian mechanics Theory (or equations) for calculating the positions of celestial bodies such as planets and stars. This theory is based on Isaac Newton's laws of motion and gravity.

open cluster A stable grouping of young stars in the *Milky Way Galaxy*. Other open clusters present in other galaxies are good indicators of spiral arms.

parallax The apparent shift in position of a nearby object projected on a background when viewed from different positions. Simply holding a finger at arm's length and viewing it with only one eye and then the other illustrates the effect. In astronomy, the parallax is the shift in position of a star when viewed from positions separated by 1 *astronomical unit*. For nearby stars, an accurate parallax fixes their distances.

parsec The distance, 3.1×10^{13} kilometers, at which a star has an astronomical parallax of 1 arcsecond. Used as a unit of measurement.

period The length of time it takes for a body to go from a starting position or condition and return to that position or condition. Period can be used to describe such actions as the rotation of a planet on its axis or its orbit around a star, and the length of time it takes a star to vary from maximum to minimum brightness and back to maximum.

planetary nebula A *nebula* which resembled a planetary disk in early telescopes. Now known to be an expanding gas cloud surrounding a hot star.

planets The major Solar System bodies in orbit around the Sun. In order of increasing distance from the Sun they are: Mercury, Venus, Earth, Mars, Jupiter, Saturn, Uranus, Neptune, and Pluto.

populations A concept used to describe different kinds of stars, which are usually distinguished by age and abundance of heavy elements. Population I stars arc young and have a high abundance of heavy elements, while Population II stars are old and have a low abundance of heavy elements.

Primordial Fireball The state of the universe that existed immediately after the *Big Bang* consisting of energetic elementary particles and radiation.

proton The heavy, positively charged elementary particle in the nucleus of the atom.

proton–proton chain Nuclear reactions that convert four hydrogen nuclei into one helium nucleus with a release of energy. This chain occurs at lower temperatures than the *carbon cycle*, and is the principal energy source for the Sun.

protoplanets Planets at an early stage of formation from clumps of gas and dust.

protostars Stars at an early stage of formation from interstellar clouds of gas and dust.

pulsar A rapidly rotating *neutron star* which emits characteristic pulses of radio radiation and visible light.

quantum mechanics The physical laws that describe the motions of electrons in atoms. These laws are radically different from *Newtonian mechanics* in that only certain electron orbits or energies are allowed. Thus, atoms emit or absorb light in discrete amounts called quanta. See *absorption line, continuous spectrum*, and *emission line*.

quasars Also called quasi-stellar objects or QSOs, quasars are starlike objects which produce emission lines. Their redshifts can be large, and their brightness varies. They are believed to be objects a little larger than the Solar System that have an intrinsic brightness some 100 times that of bright galaxies.

radiation pressure The tiny force exerted by photons when they bounce off small dust particles or when they are absorbed by atoms. The force is so small that it has no effect on the large objects encountered in everyday life, but it can be important in astronomy.

resolution See *spatial resolution, spectroscopy*, and *temporal resolution*.

RR Lyrae stars Pulsating variable stars of luminosity roughly 100 times the *intrinsic brightness* of the Sun. They are useful distance indicators, and are often found in *globular clusters*.

shock A sharp change in the properties (density, pressure, temperature) of a gas.

signal-to-noise Measurements of light in astronomy are always composed of two parts. The first is the signal, the part produced by the pure light from the target. The second is the noise, the part that comes from other sources, such as the telescope, the detectors, etc. The ratio of signal-to-noise indicates the quality of the measurement.

solar nebula The cloud of gas and dust from which the Solar System, including the Sun, formed.

solar wind The low-density gas consisting primarily of *protons* and *electrons* flowing away from the Sun at about 400 kilometers per second.

South Atlantic Anomaly A region of intense charged particle fluxes (caused by an irregularity in the Earth's magnetic field) over the south Atlantic Ocean. Spacecraft data collected in this region are often unreliable.

spatial resolution The measure of the ability of a telescope and camera to separate objects clearly. Roughly, the smallest detail that can be seen in an image.

spectral lines Because *electrons* in the cloud around the nucleus of an atom can have only a restricted number of specific energies, changes in energy produce photons of energy unique to that particular atom. This is the origin of spectral lines. These specific lines show that an element is present, and the strength of the line can be used to determine how abundant the element is. Shifts of the line in wavelength via the *Doppler effect* give the motion (speed) toward or away from the observer. The motion of atoms (with a speed depending on temperature) will cause a broadening of the line because some of the motions are away from and some are toward the observer. Thus, line widths can yield temperatures. Densities can be determined from the strengths of several different lines or from line widths. Finally, magnetic and electrical fields, if strong enough, can affect the appearance of the lines. See *absorption lines, emission lines*, and *spectroscopy*.

spectral resolution See *spectroscopy*.

spectroscopy The analysis of light to determine the properties of the medium producing it and/or influencing it. In principle, spectroscopy can yield compositions, abundances, speed (via the *Doppler effect*), temperatures, densities, and, if they are strong enough, magnetic and electric fields. The rise of modern astrophysics clearly parallels the development of spectroscopy. To carry out spectroscopy, we need to disperse, or split up, the light into its component wavelengths, or colors. This can be done by means of a prism, but in modern spectrographs the dispersive optical element is a diffraction grating, which consists of a system of precisely aligned, narrowly spaced grooves. When carrying out spectroscopy, sufficient spectral resolution is needed to separate features in the spectrum. Numerically, the spectral resolution is given by the wavelength (λ) divided by the difference in wavelength to the closest wavelength that can be separated ($\Delta\lambda$). Thus, $\lambda/\Delta\lambda$ gives us a measure of the spectral resolution of instruments, including the HST spectrographs. The concept of spectral resolution is the spectral analog of *spatial resolution* that determines the smallest detail that can be seen in ordinary images.

spiral arm The region in a spiral galaxy (such as the *Milky Way Galaxy*) that contains concentrations of gas, dust, and young stars.

stellar wind The general, steady flow of gas away from stars resulting in loss of mass. These range from the gentle *solar wind* to vigorous flows some 100 million times stronger in mass loss, such as those from Melnick 42.

supergiant A star with maximum intrinsic brightness and low density. The radius of a supergiant can be as large as 1000 times that of the Sun.

supernova An immense stellar explosion which can increase a star's *intrinsic brightness* by as much as a billion times. The explosion blows off a major fraction of the star to form an expanding gas cloud (such as the Crab Nebula). The remaining material forms a dense object such as a *neutron star* or a *black hole*.

synchrotron radiation The radiation emitted when accelerated charged particles are spiraling in a magnetic field.

temporal resolution The measure of the ability of an optical system to clearly separate events in time. Roughly, the shortest time interval that can be determined between two different events.

triple-alpha process The nuclear reactions that transform three helium nuclei (often called alpha particles) into a carbon nucleus, with a release of energy. The process occurs at high temperatures and is important in red *giant* stars.

ultraviolet dropout Light from distant galaxies can be absorbed by redshifted intergalactic hydrogen causing the image to disappear or 'dropout' in different colors, most commonly in the ultraviolet.

white dwarf An approximately Earth-sized star that does not have a source of nuclear energy in its interior. The star is supported by means of a form of pressure that arises when the densities at the star's interior are so high that the usual orbits of electrons in atoms around the nucleus cannot exist and the electrons are pushed much closer to the nucleus. A white dwarf can be supported in this way as long as its mass does not exceed about 1.4 solar masses. For stars with greater masses, *neutron stars* or *black holes* are formed.

References
and further reading

BOOKS AND PROCEEDINGS

Astronomy!, by James B. Kaler (Harper Collins, 1994).

Astronomy: From the Earth to the Universe, by Jay Pasachoff (Saunders College Publishing, 1998).

Burnham's Celestial Handbook, by Robert Burnham, Jr (Dover, 1978).

Choosing and Using a CCD Camera, by Richard Berry (Willmann-Bell, Inc., 1992).

Cosmic Landscape: Voyages Back Along the Photon's Track, by Michael Rowan-Robinson (Oxford University Press, 1979).

Cosmos, by Carl Sagan (Random House, 1980).

Electronic and Computer-Aided Astronomy, by Ian S. Mclean (Wiley and Sons, 1989).

First Light, by Richard Preston (Urania, Inc., 1987).

Light from the Depths of Time, by Rudolf Kippenhahn (Springer-Verlag, 1987).

Man Discovers the Galaxies, by Richard Berendzen, Richard Hart, and Daniel Seeley (Watson Academic Publishers, 1976).

Man Into Space, by Hermann Oberth (Harper and Brothers, 1957).

Mars, edited by H. H. Kieffer, B. M. Jakosky, C. W. Snyder, and M. S. Mathews (University of Arizona Press, 1992).

New Horizons in Astronomy (2nd edition) by J. C. Brandt and S. P. Maran (W. H. Freeman, 1979).

Science with the Hubble Space Telescope, edited by P. Benvenuti and E. J. Schreier. ESO Conference and Workshop Proceedings, no. 44, 1992.

Science with the Hubble Space Telescope II, edited by P. Benvenuti, S. D. Macchetto and E. J. Schreier. STScI–ECF Conference and Workshop Proceedings, 1996.

Stars and Planets, by Jay M. Pasachoff and Donald H. Menzel (Houghton Mifflin, 1992).

The Astronomy and Astrophysics Encyclopedia, edited by Stephen P. Maran (Cambridge University Press and van Nostrand Reinhold, 1992).

The Decade of Discovery in Astronomy and Astrophysics (National Academy Press, 1991).

The Discovery of Our Galaxy, by Charles A. Whitney (Alfred Knopf, 1971).

The First Year of HST Observations, edited by A. L. Kinney and J. C. Blades (NASA Publications, 1991).

The Hubble Wars, by Eric Chaisson (Harper Collins, 1994).

The Milky Way (5th edition), by Bart J. Bok and Priscilla F. Bok (Harvard University Press, 1981).

The New Solar System (4th edition), edited by J. Kelly Beatty, Carolyn Collins Petersen, and Andrew Chaikin (Sky Publishing and Cambridge University Press, 1998).

The Space Telescope, by Robert W. Smith (Cambridge University Press, 1993).

Voyages Through the Universe, by A. Fraknoi, D. Morrison, and S. Wolff (Saunders College Publishing, 1997).

SELECTED PAPERS

Bahcall, John N., and Spitzer, Lyman. The Space Telescope, *Scientific American*, **247**, no. 1, July 1982.

Brandt, J.C. *et al.* The Goddard High Resolution Spectrograph: Instrument, Goals and Science Results, *Publications of the Astronomical Society of the Pacific*, **106**, 702, August 1994.

Elliott, J.L. *et al.* An Occultation by Saturn's rings on October 2-3 1991. Observed with the Hubble Space Telescope, *Astrophysical Journal*, Dec. 1993.

Hensen, B.W. ESA's First In-Orbit-Replaceable Solar Array, *ESA Bulletin No. 61*, February 1990.

Linsky, J.L. Ultraviolet Observations of Stellar Coronae: Early Results from HST, *Memorie della Societa Astronomica Italiana*, **63**, no. 3, 1992.

Spitzer, Lyman. Astronomical Advantages of an Extra-Terrestrial Observatory, *The Astronomical Quarterly*, **7**, 131–42, 1990. (Reprint of Appendix V of Spitzer's 1946 RAND Report.)

OTHER ARTICLES AND PUBLICATIONS OF INTEREST

Exploring the Universe with the Hubble Space Telescope, NASA NP-126.

Scientific Uses of the Large Space Telescope, National Academy of Sciences, Washington, D.C., 1969 (Committee Report, Lyman Spitzer, Chairman).

Time, Money and Millionths of an Inch, *Hartford Courant*, March 31–April 3, 1991.

Interested readers should also consult *The Space Telecope Science Institute Newsletter*, published quarterly by the Space Telescope Science Institute.

HST results are often reported in the following magazines and journals:

> *Astronomical Journal*
> *Astronomy and Astrophysics*
> *Astronomy Magazine*

Astronomy Now
Astrophysical Journal
Bulletin of the American Astronomical Society
Mercury Magazine
New Scientist
Physics Today
Scientific American
Sky & Telescope Magazine

In addition, up-to-date reports on Hubble science can be found on the World Wide Web on the following sites, each with links to other useful sites:

The Space Telescope Science Institute
http://www.stsci.edu

STScI Office of Public Outreach
http://oposite.stsci.edu/pubinfo/

Index